AF730474

TECHNICAL MANUAL
No. 5-725

HEADQUARTERS
DEPARTMENT OF THE ARMY
WASHINGTON, D.C., 3 October 1968

RIGGING

*This manual supersedes TM 5-725, 20 March 1964.

This page intentionally left blank

CHAPTER 1
INTRODUCTION

Section I. GENERAL

1–1. Purpose and Scope

a. This manual is a guide and basic reference for personnel whose duties require the use of rigging. It is intended for use in training and as a handbook for field operations. It covers the types of rigging and the application of fiber rope, wire rope, and chains used in various combinations to raise or move heavy loads. It includes basic instruction on knots, hitches, splices, lashing, and tackle systems. Safety precautions and requirements for the various operations are listed, as well as rules of thumb for rapid safe-load calculation.

b. The material contained herein is applicable to both nuclear and nonnuclear warfare.

1–2. Changes

Users of this manual are encouraged to submit recommended changes or comments to improve this manual. Comments should be keyed to the specific page, paragraph, and line of the text in which the change is recommended. Reasons should be provided for each comment to insure understanding and complete evaluation. Comments should be forwarded direct to Commandant, United States Army Engineer School, Fort Belvoir, Va. 22060.

Section II. FIBER ROPE

1–3. Kinds

The term cordage is applied collectively to ropes and twines made by twisting together vegetable or synthetic fibers. The principal vegetable fibers are abacá, sisalana, henequen, hemp, and sometimes cotton and jute. The last two are relatively unimportant in the heavy cordage field. Abacá (known as Manila), sisalana and henequen (both known as sisal), are classed as hard fibers. The comparative strengths of the above fibers, considering abacá as 100, are as follows: sisalana 80, henequen 65, and hemp 100.

a. Manila. Manila is a strong fiber that comes from the leaf stems of the stalk of the abacá plant, which belongs to the banana family. The fibers vary in length from 1.2 to 4.5 meters (4 to 15 feet) in the natural states. The quality of the fiber and its length give manila rope relatively high elasticity, strength, and resistance to wear and deterioration. In many instances the manufacturer treats the rope with chemicals to make it more mildew resistant, which increases the quality of the rope. Manila rope is generally the standard item of issue because of its quality and relative strength.

b. Sisal. Sisal rope is made from two tropical plants that yield a strong, valuable fiber. These plants, sisalana and henequen, produce fibers 0.6 to 1.2 meters (2 to 4 feet) long with sisalana producing the stronger fibers of the two plants. Because of the greater strength of sisalana, these fibers are used to make the rope known as sisal. Sisal rope is about 80 percent as strong as high quality manila rope and can be easily obtained. It withstands exposure to sea water very well and is often used for this reason.

c. Hemp. Hemp is a tall plant that provides useful fibers for making rope and cloth. It is cultivated in many parts of the world. Hemp was used extensively before the introduction of

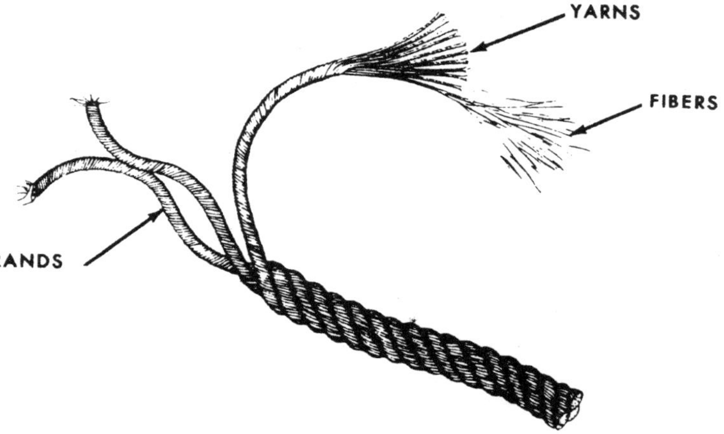

Figure 1-1. Cordage elements of rope construction.

manila, but its use today is relatively small. Its principal use now is in fittings such as ratline, marline, and spun yarn. Since hemp absorbs far much better than the hard fibers, these fittings are invariably tarred to make them more water resistant. Tarred hemp has about 80 percent of the strength of untarred hemp. Of these tarred fittings, marline is the standard item of issue.

d. Coir and Cotton. Coir rope is made from the fiber of coconut husks. It is a very elastic rough rope about one-fourth the strength of hemp, but light enough to float on water. Cotton makes a very smooth white rope which stands much bending and running. These two types of rope are not widely used in the military service except that cotton is used in some cases for very small lines.

e. Nylon. Nylon has a tensile strength nearly three times that of manila. The advantage of using nylon rope is that it is waterproof and has the ability to stretch, absorb shocks and resume normal length. It also resists abrasion, rot, decay, and fungus growth.

1-4. Fabrication

In the fabrication of fiber rope (fig. 1-1), a number of fibers of various plants are twisted together to form the yarns, which are then twisted together in an opposite direction to the fibers to form the strands. The strands are twisted in an opposite direction to the yarns to form the completed rope. The direction of twist of each element of the rope is known as the "lay" of that element. Twisting each element in the opposite direction (fig. 1-1) puts the rope in balance and prevents its elements from unlaying when a load is suspended on it. The principal type of rope is the three strand, right lay, in which three strands are twisted in a right-hand direction. Four strand ropes, which are also available, are slightly heavier but weaker than three-strand ropes of the same diameter.

1-5. Characteristics

a. Size. In the U.S. Army, wire and fiber rope sizes are designated by inches of diameter and circumference. Wire rope is always designated by diameter; however, fiber rope is designated by diameter up to 5/8 inch. Then it is designated by circumference up to 12 inches or more. For this reason most tables give both diameter and circumference of fiber rope.

Table 1-1. Properties of Manila and Sisal Rope

Nominal diameter (inches)	Circumference (inches)	Lb per ft	No. 1 Manila: Breaking strength (pounds)	No. 1 Manila: Safe load (pounds) F.S. = 4	Sisal: Breaking strength (pounds)	Sisal: Safe load (pounds) F.S. = 4
¼	¾	.020	600	150	480	120
⅜	1⅛	.040	1,350	325	1,080	260
½	1½	.075	2,650	660	2,120	520
⅝	2	.133	4,400	1,100	3,520	880
¾	2¼	.167	5,400	1,350	4,320	1,080
⅞	2¾	.186	7,700	1,920	6,160	1,540
1	3	.270	9,000	2,250	7,200	1,800
1⅛	3½	.360	12,000	3,000	9,600	2,400
1¼	3¾	.418	13,500	3,380	10,800	2,700
1½	4½	.600	18,500	4,620	14,800	3,700
1¾	5½	.895	26,500	6,625	21,200	5,300
2	6	1.08	31,000	7,750	24,800	6,200
2½	7½	1.35	46,500	11,620	37,200	9,300
3	9	2.42	64,000	16,000	51,200	12,800

NOTES:

1. Breaking strengths and safe loads given are for new rope used under favorable conditions. As rope ages or deteriorates, progressively reduce safe loads to one-half of values given.

2. Safe working load may be computed using a safety factor of 4, but when the condition of the rope is doubtful, further divide the computed load by 2.

b. Weight. The weight of rope varies with use, weather conditions, added preservatives, and other factors. Table 1-1 lists the weight of new fiber rope.

c. Strength. Table 1-1 lists some of the properties of manila and sisal rope, including strength. The table shows that the minimum breaking strength is considerably greater than the safe working capacity. The difference is caused by the application of a safety factor. The safe working capacity of rope is obtained by dividing the breaking strength by a factor of safety (F.S.) $\left(SWS = \frac{BS}{FS} \right)$. A new 1 inch diameter No. 1 Manila rope has a breaking strength of 9,000 pounds (table 1-1). To determine the rope's safe working capacity, divide its breaking strength (9,000 pounds) by a minimum standard safety factor of 4. The result is a safe working capacity of 2,250 pounds. This means that we can safely apply 2,250 pounds of tension to the new one-inch diameter No. 1 Manila rope in normal use. A safety factor is always used because the breaking strength of rope becomes reduced after use and exposure to weather conditons. In addition, a safety factor is required because of shock loading, knots, sharp bends and other stresses which the rope may have to withstand during its use. Some of these stresses reduce the strength of rope as much as 50 percent. If tables are not available, the safe working capacity may be closely approximated by a rule of thumb. The rule of thumb for the safe working capacity, in tons, for fiber rope is equal to the square of the rope diameter in inches ($SWC = D^2$). The safe working capacity, in tons, of a ½-inch diameter fiber rope would be ½ inch squared or ¼ ton. The rule of thumb will allow a safety factor of approximately 4.

1-6. Care

The strength and useful life of fiber rope will be shortened considerably by improper care. Fiber rope should be dry when stored and should be stored in a cool, dry place. This reduces the chances of mildew and rotting. It should be coiled on a spool or hung from pegs in a way that will allow circulation of air. Avoid dragging the rope through sand or dirt, or pulling the rope over sharp edges. Sand or grit between the fibers of the rope will cut the fibers and reduce its strength. Slacken taut lines before they are exposed to rain or dampness because a wet rope shrinks and may break. A frozen rope should not be used until it is completely thawed; otherwise the frozen fibers will be broken as they resist bending. Avoid exposure of fiber rope to excessive heat and fumes of chemicals. Exposure to heat or

UNCOILING A NEW COIL OF FIBER ROPE

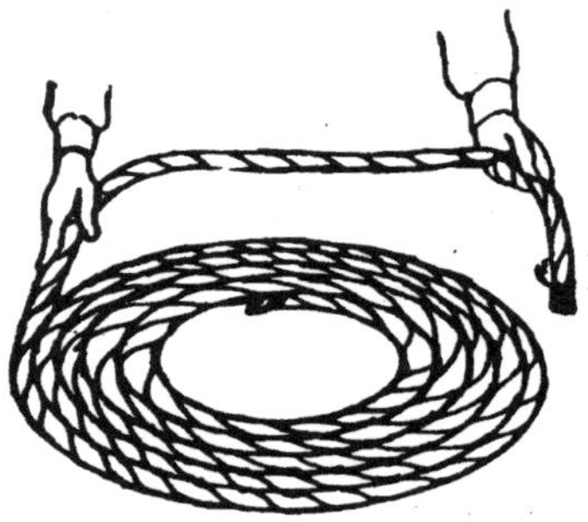

COILING OF A FIBER ROPE AFTER BEING UTILIZED

Figure 1–2. Uncoiling and coiling rope.

boiling water will decrease rope strength about 20 percent.

1–7. Handling

New rope is coiled, bound, and wrapped in burlap. The protective covering should not be removed until the rope is to be used. This will protect the rope during storage and prevent tangling. To open, strip off the burlap wrapping and look inside the coil for the end of the rope. This should be at the bottom of the coil (fig. 1-2). If it is not, turn the coil over so that the end will be at the bottom. Pull the end of the rope up through the center of the coil; as the rope comes up through the coil it will unwind in a counterclockwise direction.

1-8. Inspection

The outside appearance of fiber rope is not always a good indication of its internal condition. The rope will soften with use. Dampness, heavy strain, the fraying and breaking of strands, and chafing on rough edges, all weaken the rope considerably. Overloading of a rope may cause it to break, with possible heavy damage to material and serious injury to personnel. For this reason, inspect rope carefully at regular intervals to determine its condition. Untwist the strands slightly to open the rope so that the inside can be examined. Mildewed rope will have a musty odor and inner fibers of the strands will have a dark stained appearance. Broken strands or broken yarns ordinarily are easy to identify. Dirt and sawdust-like material inside the rope, caused by chafing, indicate damage. In rope having a central core, the core should not break away in small pieces upon examination. If this happens, it indicates that the rope has been overstrained. If the rope appears to be satisfactory in all other respects, pull out two fibers and try to break them. Sound fibers should offer considerable resistance to breakage. When any unsatisfactory conditions are found, destroy the rope or cut it up in short pieces. Make sure none of these pieces is left long enough to permit its use in hoisting. This prevents the use of the rope for hoisting, but saves the short pieces for miscellaneous use.

Section III. WIRE ROPE

1-9. Fabrication

The basic element of wire rope is the individual wire, which is made of steel or iron in various sizes. The wires are laid together to form strands. The strands are laid together to form the rope (fig. 1-3). The individual wires are usually wound or laid together in a direction opposite to the lay of the strands. The strands are then wound about a central core which supports and maintains the position of strands during bending and load stresses. The core may be constructed of fiber rope, independent wire rope, or wire strand. The fiber core can be either vegetable or synthetic fiber rope. This type of wire rope has the fiber core treated with a special lubricant which helps keep the wire rope lubricated internally. Under tension, the wire rope contracts, forcing the lubricant from the core into the rope. This type of core has the additional advantage of acting as a cushion for the strands when under stress. After contracting, the fiber core acts as a stress absorbent and prevents the internal crushing of the individual wires. The limitations of fiber cores are reached when pressure, such as crushing on the drum, results in core collapse and distortion of the rope strand. Furthermore, if the rope is subjected to excessive heat, the vegetable or synthetic fibers may be damaged. Under such severe conditions, independent wire rope cores are normally used. The independent wire rope core is actually a separate smaller wire rope acting as the core. The independent wire rope core also adds to the strength of the rope. A wire strand core consists of a single strand either of the same construction or sometimes more flexible than the main rope strands. In some wire ropes, the wires and strands are preformed. Preforming is a method of presetting the wires in the strands (and the strands in the rope) into the permanent helical or corkscrew form they will have in the completed rope. As a result, preformed wire rope doesn't contain the internal stresses found in the nonpreformed wire rope; therefore, it does not untwist as easily and is more flexible than nonpreformed wire rope.

1-10. Classification

Wire rope is classified by the number of strands, number of wires per strand, strand construction, and type of lay.

a. Combination. Wire and strand combinations (fig. 1-4) vary according to the purpose for which the rope is intended. The smaller and more numerous the wires the more flexible the rope, but the less resistant to external abrasion. Rope made up of a smaller number of larger wires is more resistant to external abrasion but is less flexible. The 6 x 37 (six

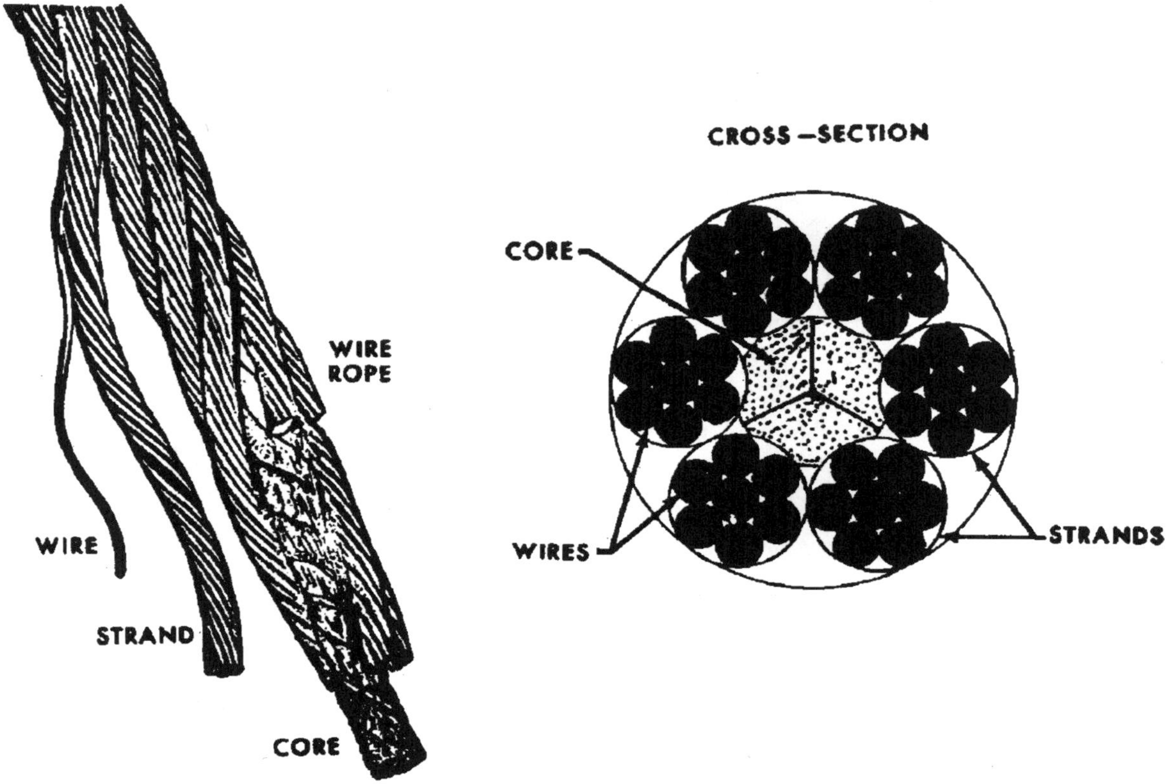

Figure 1–3. Elements of construction of wire rope.

strands, each made up of 37 wires) wire rope is the most flexible of the standard six-strand ropes. This permits its use with small sheaves and drums, such as on cranes. It is a very efficient rope because many inner strands are protected from abrasion by the outer strands. The stiffest and strongest type for general use is the 6 x 19 rope. This rope may be used over sheaves of large diameter if the speed is kept to moderate levels. It is not suitable for rapid operation or for use over small sheaves because of its stiffness. Wire rope 6 x 7 is the least flexible of the standard rope constructions. It is well suited to withstand abrasive wear because of the large outer wires.

b. Lay. Lay (fig. 1–5) refers to the direction of winding of the wires in the strands and the strands in the rope. Both may be wound in the same direction, or they may be wound in opposite directions. There are three types of rope lays:

(1) *Regular lay*. In regular lay, the strands and wires are wound in opposite directions. The most common lay in wire rope is the right regular lay. (Strands wound right, wires wound left.) Left regular lay (strands wound left, wires wound right) is used where the untwisting rotation of the rope will counteract the unscrewing forces in the supported load as, for example, in drill rods and tubes for deep well drilling.

(2) *Lang lay*. In Lang lay, the strands and wires are wound in the same direction. Because of the greater length of exposed wires, the Lang lay assures longer abrasion resistance of the wires, less radial pressure on small diameter sheaves or drums by the ropes and less binding stresses in wire than in regular lay wire rope. Disadvantages of the Lang lay are the tendency to kinking and unlaying or open-up of the strands, which makes it undesirable for use where grit, dust, and moisture are present. The standard direction of Lang lay is right, (strands and wires wound right) al-

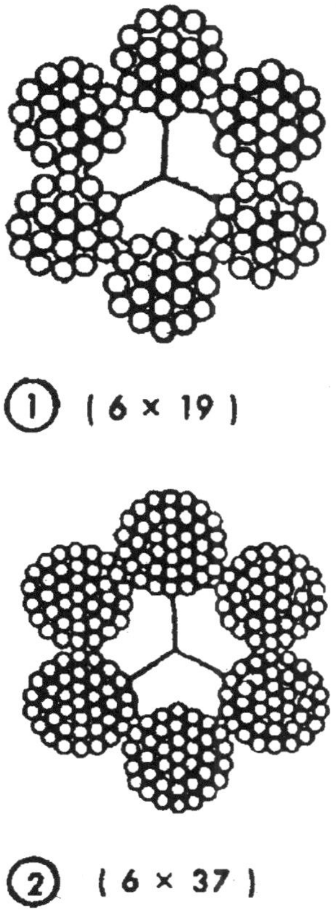

Figure 1–4. Arrangement of strands in wire rope.

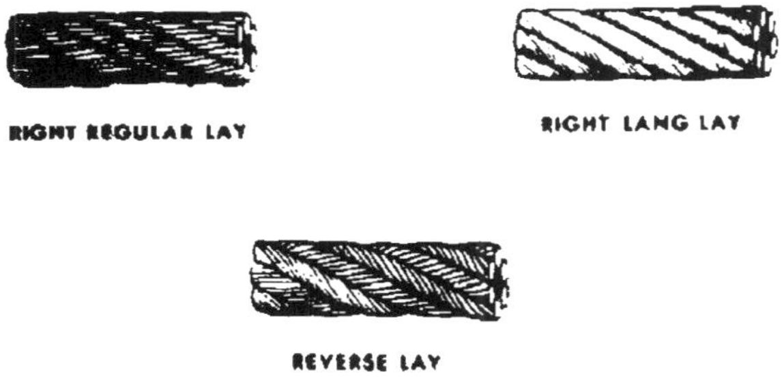

Figure 1–5. Wire rope lays.

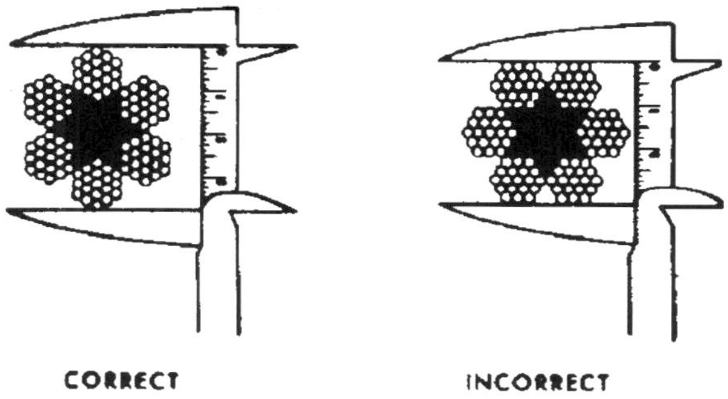

Figure 1–6. Measuring wire rope.

Figure 1–7. Kinking in wire rope.

though it also comes in left lay (strands and wires wound left.)

(3) *Reverse lay.* In reverse lay, the wires of any strand are wound in the opposite direction of the wires in the adjacent strands. Reverse lay applies to ropes in which the strands are alternately regular and Lang lay. The use

Figure 1–8. Unreeling wire rope.

of reverse lay rope is usually limited to certain types of conveyors. The standard direction of lay is right, (strands wound right) as it is for both regular lay and Lang lay ropes.

1–11. Characteristics

a. Size. The size of wire rope is designated by its diameter in inches. To determine the size of a wire rope, measure its greatest diameter (fig. 1–6).

b. Weight. The weight of wire rope varies with the size and the type of construction. No rule of thumb can be given for determining the weight. Approximate weights for certain sizes are given in table 1–2.

c. Strength. The strength of a wire rope is determined by its size, grade, and method of fabrication. The individual wires may be made of various materials including traction steel, mild plow steel, improved plow steel, and extra improved plow steel. The ultimate or maximum strength of a wire rope is referred to as the breaking strength. Since a suitable margin of safety must be provided when applying a load to a wire rope, the breaking strength is divided by an appropriate safety factor (table 1–3).

Table 1–2. Breaking Strength of 6 x 19 Standard Wire Rope [1]

Diameter n.	Approximate weight lb/ft	Iron	Breaking strength, tons of 2000 lbs [2]			
			Traction steel	Plow steel	Improved plow steel	Extra improved plow steel
1½	3.60	29.7	56.6	108.92	123.43	123.43
¼	0.10	1.4	2.6	2.39	2.74	---
⅜	0.23	2.1	4.0	5.31	6.10	7.55
½	0.40	3.6	6.8	9.35	10.7	13.3
⅝	0.63	5.5	10.4	14.5	16.7	20.6
¾	0.90	7.9	14.8	20.7	23.8	29.4
⅞	1.23	10.6	20.2	28.0	32.2	39.8
1	1.60	13.7	26.0	36.4	41.8	51.7
1⅛	2.03	17.2	32.7	45.7	52.6	65.0
1¼	2.50	21.0	40.6	56.2	64.6	79.9
1½	3.60	29.7	56.6	80.0	92.0	114.0
1¾				108.0	124.0	153.0
2				139.0	160.0	198.0

[1] 6 x 19 means rope composed of 6 strands of 19 wires each.

[2] The maximum allowable working load is the breaking strength divided by the appropriate factor of safety. See Table 1–3.

Figure 1–9. Uncoiling wire rope.

The value then obtained is referred to as the safe working capacity and is the maximum load that can safely be applied to the rope for that particular type of service. The factors of safety given in table 1–3 should be used in all cases where the rope will be in service for a considerable time. As a rule of thumb the diameter of wire rope in inches can be squared and multiplied by 8 to obtain the safe working capacity in tons. (SWC = $8D^2$) A value obtained in this manner will not always agree with the factor of safety given in table 1–3. The table is more accurate. The proper safety factor depends not only on loads applied, but also on the speed of operation, the type of fittings used for securing the rope ends, the acceleration and deceleration, the length of rope, the number, size and location of sheaves and drums, the factors causing abrasion and corrosion, the facilities for inspection, and the possible loss of life and property should a rope fail. Table 1–2 shows comparative breaking strengths of typical wire ropes.

Table 1–3. Wire Rope Safety Factors

Type of service	Minimum safety factor
Track cables	3.2
Guys	3.5
Miscellaneous hoisting equipment	5.0
Haulage ropes	6.0
Derricks	6.0
Small electric and air hoists	7.0
Slings	8.0

1–12. Care

a. Lubrication. At the time of fabrication a lubricant is applied to wire rope. This lubricant generally does not last throughout the life of the rope, which make relubrication necessary. A good grade of oil or grease can be used for this purpose; it should be free of acids and alkalies and should be light enough to penetrate between the wires and strands of the rope. The lubricant can be brushed on the rope, or it can be applied by passing the rope through a trough or box containing the lubricant. The lubricant should be applied as uniformly as possible throughout the length of the rope. In every case where the wire rope is being stored for any length of time it should be cleaned and lubricated before storage.

b. Cleaning. Scraping or steaming will remove most of the dirt or grit which may have accumulated on a used wire rope. Rust should be removed at regular intervals by wire brushing. The rope should always be carefully cleaned just before lubrication. The object of cleaning at that time is to remove all foreign material and old lubricant from the valleys between the strands and from the spaces between the outer wires to permit the newly applied lubricant free entrance into the rope.

c. Reversing Ends. To obtain increased service from wire rope it is sometimes advisable to reverse ends or cut back the ends. Reversing ends is more satisfactory than just cutting back the ends because frequently the wear and fatigue on a rope are more severe at certain points than at others. To reverse ends, the drum end of the rope is detached from the drum and placed in the end attachment. The end removed from the end attachment then is fastened to the drum. Cutting back the end has a similar effect, but there is not as much change involved. A short length is cut off the end of the rope and the new end placed in the fitting, thus removing the section which has sustained the greatest local fatigue.

d. Storage. Wire rope should be coiled on a spool for storage and should be properly tagged as to size and length. It should be stored in a dry place to reduce corrosion and kept away from chemicals and fumes which might attack the metal. Before storage, wire rope should always be cleaned and lubricated. If the lubricant film is applied properly and the wire is stored in a place protected from the weather, corrosion will be virtually eliminated. Rusting, corrosion of the wires, and deterioration of the fiber core sharply decrease the strength of the rope. The loss of strength caused by these effects is difficult to estimate.

1–13. Handling

a. Kinking. When loose wire rope is handled, small loops (fig. 1–7) frequently form in the slack portion of the rope. If tension is applied to the rope while these loops are in position, they will not straighten out but will form sharp kinks, resulting in unlaying of the rope. All of these loops should be straightened out of the rope before applying a load. After a kink has formed in wire rope it is impossible to remove it, and the strength of the rope is seriously damaged at the point where the kink occurs. Such a kinked portion should be cut out of the rope before it is used again.

b. Coiling. Small loops or twists will form if the rope is being wound into the coil in a direction opposite to the lay of the rope. Left lay wire rope should be coiled in a counterclockwise direction and right lay wire rope should be coiled in a clockwise direction.

c. Unreeling. When removing wire rope from a reel or coil, it is imperative that the reel or coil rotate as the rope unwinds. The reel may be mounted as shown in figure 1–8. The rope is then pulled from the reel by a man holding the end of the rope and walking away from the reel which rotates as the rope unwinds. If a wire rope is in a small coil, stand the coil on end and roll it along the ground (fig. 1–9). If loops form in the wire rope, they should be carefully removed before they form kinks. A kink can severely damage wire rope.

d. Seizing. Seizing is the most satisfactory method of binding the end of a wire rope, although welding will also hold the ends together satisfactorily. The seizing will last as long as desired, and there is no danger of weakening the wire through the application of heat. Wire rope is seized as shown in figure 1–10. There are three convenient rules for determining the number of seizings, lengths, and space between seizings. In each case when the calculation results in a fraction, the next larger whole number is used. The following calculations are based on a ¾-inch diameter wire rope.

(1) The number of seizings to be applied equals approximately three times the diameter of the rope (No. seizings = 3D)

Example: 3 × ¾ (dia) = 2¼. Use 3 seizings.

(2) Each seizing should be 1 to 1½ times as long as the diameter of the rope. (Length of seizing = 1½D)

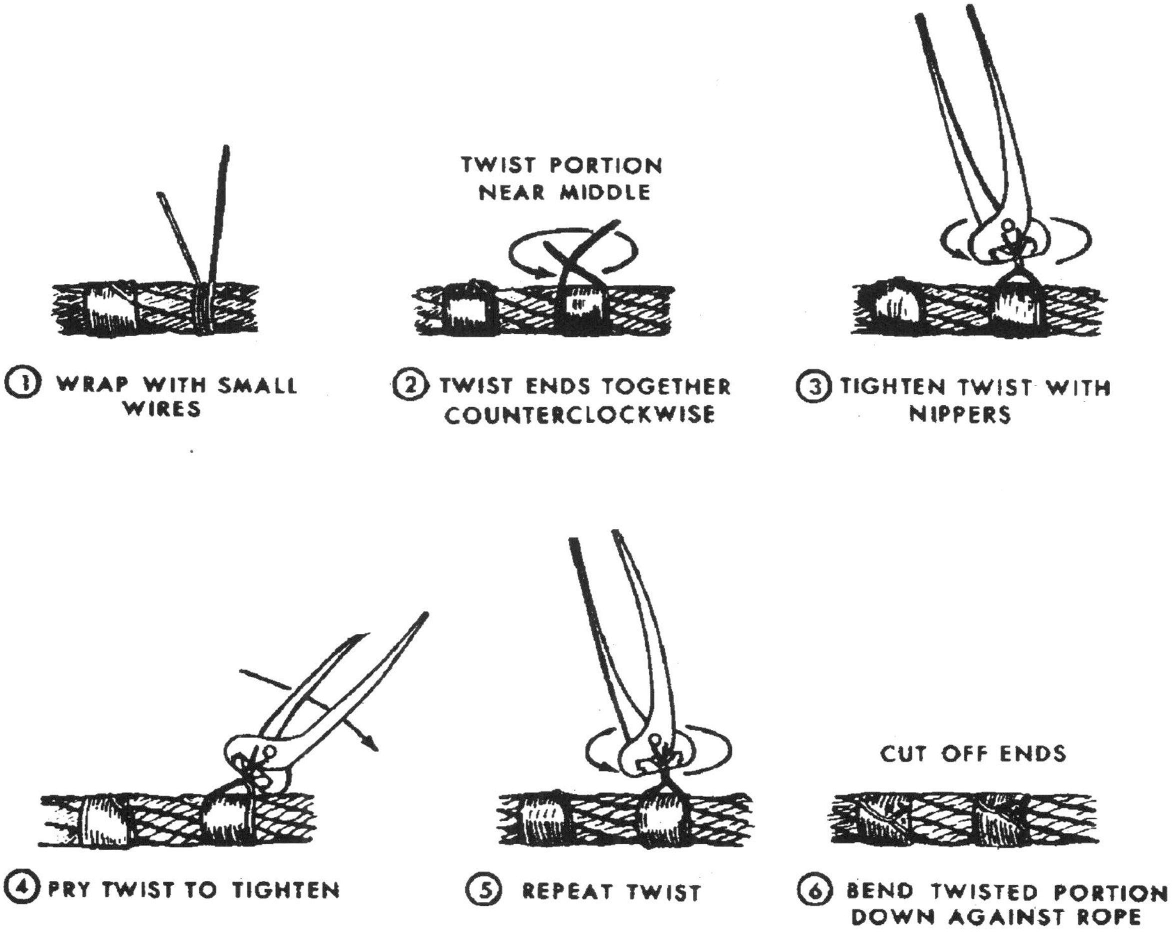

Figure 1 10. Seizing wire rope.

Example: 1 × ¾ (dia) = ¾. Use 1-inch seizings.

(3) The seizings should be spaced a distance apart equal to twice the diameter. (Spacing = 2D)

Example: 2 × ¾ (dia) = 1½. Use 2-inch spaces.

Note. Always change fraction to next larger whole number.

e. Welding. Wire rope ends may be bound together by fusing or welding the wires. This is a satisfactory method if carefully done, as it does not increase the size of the rope, and requires little time to complete. Before welding the rope a short piece of the core should be cut out of the end so that a clean weld will result and the core will not be burned deep into the rope. The area heated should be kept to a minimum, and no more heat should be applied than essential to fuse the metal.

f. Cutting. Wire rope may be cut with a wire rope cutter (fig. 1–11), a cold chisel, a hacksaw, bolt clippers, or an oxyacetylene cutting torch. Before the wire rope is cut, the strands must be tightly bound to prevent unlaying of the rope. Seizing or welding will secure the ends that are to be cut. To use the wire rope cutter, insert the wire rope in the bottom of the cutter with the blade of the cutter coming between the two central seizings. Push the blade down against the wire rope and strike the top of the blade sharply with a sledge ham-

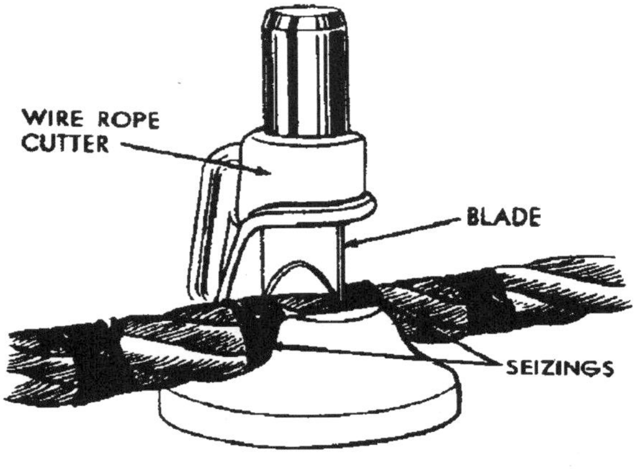

Figure 1–11. Wire rope cutter.

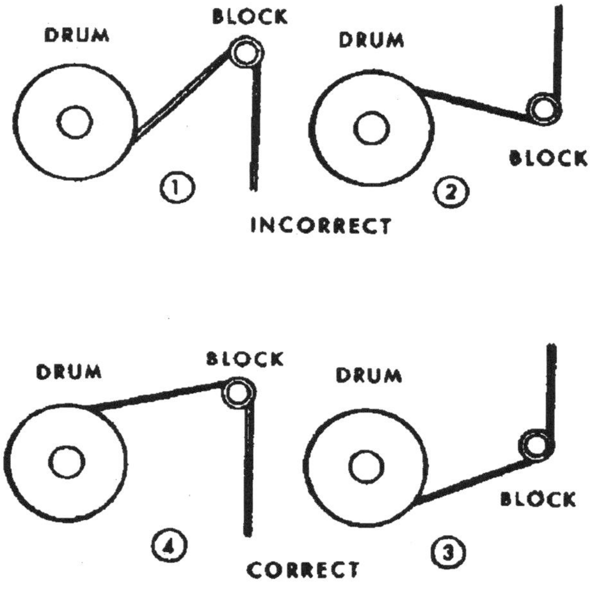

Figure 1–12. Avoid reverse bends in wire rope.

mer several times. The bolt clippers can be used on wire rope of fairly small diameter, but the oxyacetylene torch can be used on wire rope of any diameter. The hacksaw and cold chisel are slower methods of cutting.

g. Drums and Sheaves.

(1) *Size.* The size and location of the sheaves and drums about which wire rope operates and the speed with which the rope passes over the sheaves have a definite effect on the strength and service life of the rope. Each time a wire rope is bent, the individual

Figure 1–13. Spooling wire rope from reel to drum.

strands must move with respect to each other in addition to bending. This bending and moving of wires should be kept to a minimum to reduce wear. If the sheave or drum diameter is sufficiently large, the loss of strength due to bending wire rope around it will be in the neighborhood of 5 or 6 percent. In all cases, the speed of the rope over the sheaves or drum should be kept as slow as is consistent with efficient work, to decrease wear on the rope. It is impossible to give an absolute minimum size for each sheave or drum, since a number of factors enter into this decision. However, table 1–4 shows the minimum recommended sheave and drum diameters for several wire rope sizes. The sheave diameter always should be as large as possible and, except for very flexible rope, never less than 20 times the wire rope diameter. This figure has been adopted widely.

Table 1–4. Minimum Tread Diameter of Sheaves and Drums

Rope diameter in inches	Minimum tread diameter in inches for given rope construction*			
	6 x 7	6 x 19	6 x 37	8 x 19
¼	10½	8½		6½
⅜	15¾	12¾	6¾	9¾
½	21	17	9	13
⅝	26¼	21¼	11¼	16¼
¾	31½	25½	13½	19½
⅞	36¾	29¾	15¾	22¾
1	42	34	18	26
1⅛	47¼	38¼	20¼	29¼
1¼	52½	42½	22½	32½
1½	63	51	27	39

*Rope construction is strands and wires per strand.

(2) *Location.* Drums, sheaves, and blocks used with wire rope should be reeved and placed in a manner to avoid reverse bends (fig. 1–12) whenever possible. A reverse bend occurs when the rope bends in one direction around one block, drum, or sheave and bends in the opposite direction around the next. This causes the individual wires and strands to do an unnecessary amount of shifting which in-

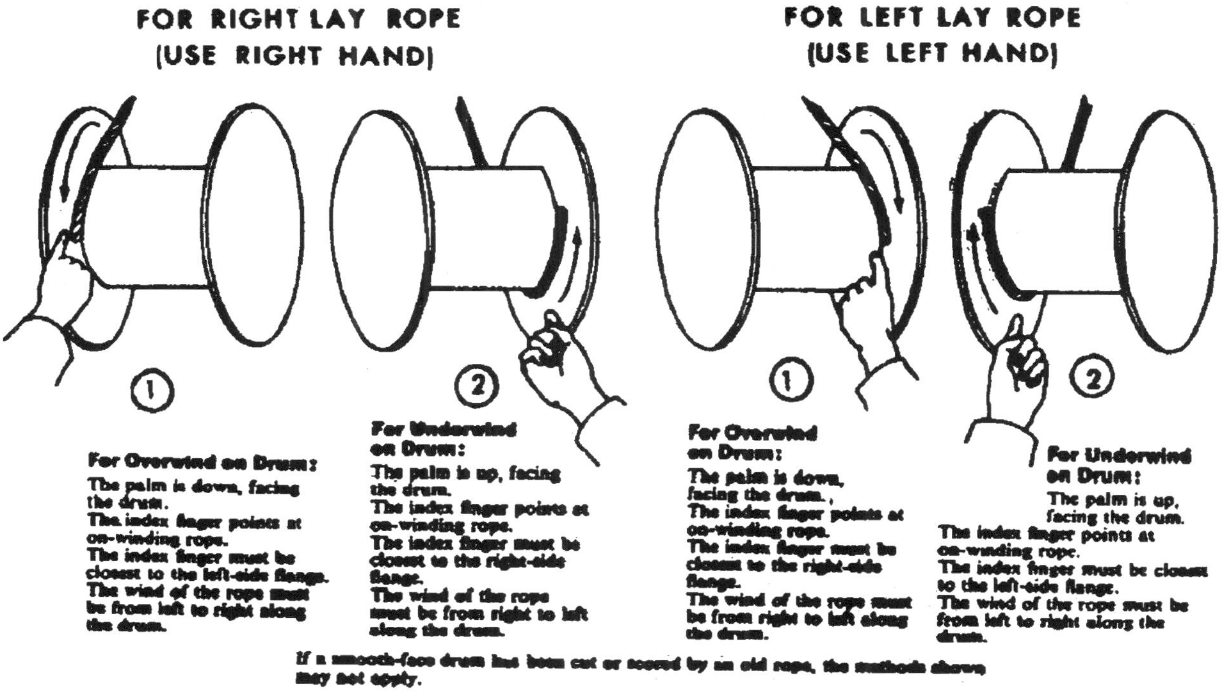

Figure 1–14. Hand rule for determining proper starting flange for wire rope.

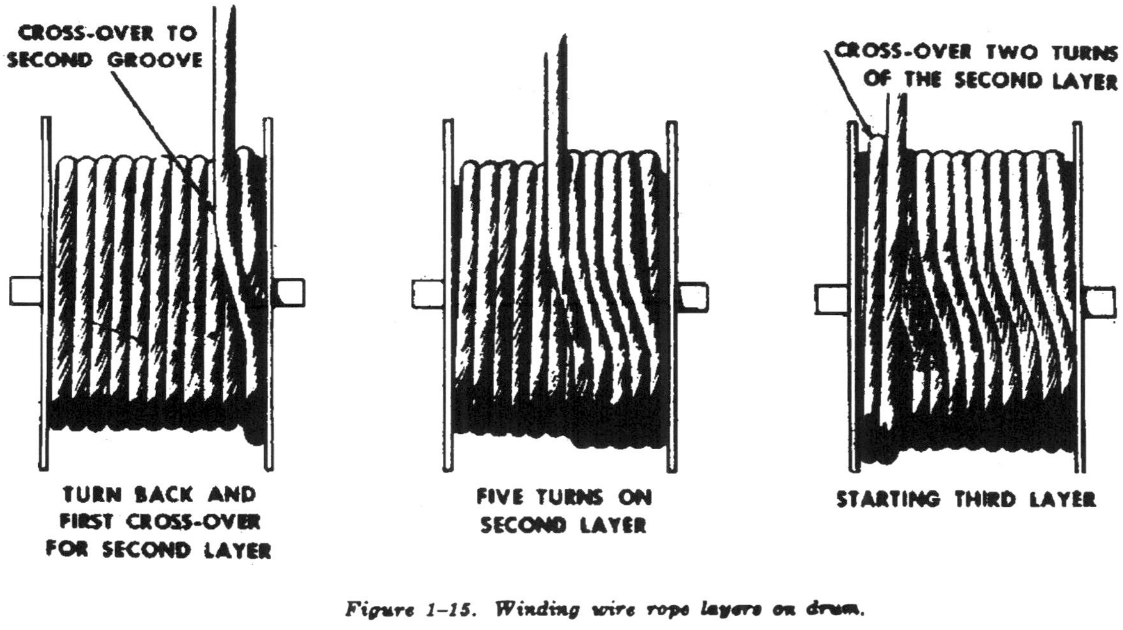

Figure 1–15. Winding wire rope layers on drum.

creases wear. Where a reverse bend must be used, the blocks, sheaves, or drum causing the reversal should be of larger diameter than ordinarily used and should be spaced as far apart as possible so there will be more time allowed between the bending motions.

(3) *Winding.* Turns of wire rope should not overlap when wound on the drum of a winch, but should be wrapped in smooth layers. Overlapping will result in binding, causing snatches on the line when the rope is unwound. To produce smooth layers, start the rope against one flange of the drum and keep a tension on the line while winding. Start the rope against the right or left flange as necessary to match the direction of winding, so that when rewound on the drum the rope will curve in the same manner as when it left the reel (fig. 1–13). A convenient method for determining the proper flange of the drum for starting the rope is known as the hand rule (fig. 1–14). The extended index finger on this figure points at the on-winding rope. The turns of rope are wound on the drum close together to prevent the possibility of crushing and abrasion of the rope while winding, and binding or snatching of the rope when it is unwound. If necessary, a wood stick should be used to force the turns closer together. Striking the wire with a hammer or other metal object damages the individual wires in the rope. If possible, only a single layer of wire rope should be wound on the drum. Where it is necessary to wind additional layers, they must be wound so that binding will be eliminated. The second layer of turns is wound over the first layer by placing the wire in the grooves formed by the first layer, except that each turn of the rope in the second layer is crossed over two turns of the first layer (fig. 1–15). The third layer is wound in the grooves of the second layer, except that each turn of the rope will cross over two turns of the second layer.

1–14. Inspection

a. Frequency. Wire ropes should be inspected frequently. Frayed, kinked, worn, or corroded ropes must be replaced. The frequency of inspection is determined by the amount of use of the rope. A rope that is used 1 or 2 hours a week requires less frequent inspection than a rope which is used 24 hours a day.

b. Procedure. The weak points in the rope and the points where the greatest stress occurs must be inspected carefully.

(1) Worn spots will show up as shiny flattened spots on the wires. If the outer wires have been reduced in diameter by one-fourth, the worn spot is unsafe.

(2) Broken wires must be inspected to determine whether it is a single broken wire or several.

(*a*) If individual wires are broken next to one another, unequal load distribution at this point will make the rope unsafe.

(*b*) When 4 percent of the total number of wires composing a type of wire rope are found to be broken in one strand (the distance in which one strand makes one complete turn around the rope), the rope is unsafe.

(*c*) The rope is unsafe if three broken wires are found in one strand of 6 x 7 rope, six broken wires are found in one strand of 6 x 19 rope, or nine broken wires are found in one strand of 6 x 37 rope.

c. Common Causes of Wire Rope Failures. There are many forms of abuse of wire ropes. The most common abuses are the use of a rope which is—

(1) Of incorrect size, construction, or grade.

(2) Allowed to drag over obstacles.

(3) Not properly lubricated.

(4) Operating over sheaves and drums of inadequate size.

(5) Overwinding or crosswinding on drums.

(6) Operating over sheaves and drums out of alinement.

(7) Permitted to jump sheaves.

(8) Subjected to moisture or acid fumes.

(9) Permitted to untwist.

(10) Kinked.

CHAPTER 2

KNOTS, SPLICES, AND ATTACHMENTS

Section I. KNOTS, HITCHES, AND LASHINGS

2–1. Introduction

A good knot must be easy to tie, hold without slipping, and be easy to untie. The choice of the best knot, bend, or hitch to use depends largely on the job it has to do. In general, knots can be classified into three groups—knots at the end of a rope, knots for joining two ropes, and knots for making loops. A study of the terminology pictured in figure 2–1 and the following definitions will aid in understanding the methods of knotting presented in this section.

a. Fundamental Terms.

(1) *Rope.* A rope (often called a line) is a large, stout cord made of strands of fiber or wire twisted or braided together.

(2) *Line.* A line (sometimes called a rope) is a thread, string, cord, or rope, especially a comparatively slender and strong cord. This manual will use the word rope rather than line in describing knots, hitches, rigging, and the like.

(3) *Running end.* The running end is the free or working end of a rope.

(4) *Standing part.* The standing part is the rest of the rope, excluding the running end.

(5) *Bight.* A bight is a bend or U-shaped curve in a rope.

(6) *Loop.* A loop is formed by crossing the running end cover or under the standing part forming a ring or circle in the rope.

(7) *Turn.* A turn is the placing of a loop around a specific object such as a post, rail, or ring with the running end continuing in a direction opposite to the standing part.

(8) *Round turn.* A round turn is a modified turn, but with the running end leaving the circle in the same general direction as the standing part.

(9) *Overhand turn or loop.* An overhand turn or loop is made when the running end passes over the standing part.

(10) *Underhand turn or loop.* An underhand turn or loop is made when the running end passes under the standing part.

(11) *Knot.* A knot is an interlacement of the parts of one or more flexible bodies, as cordage rope, forming a lump known as a knot; any tie or fastening formed with a rope, including bends, hitches, and splices. It is often used as a stopper to prevent a rope from passing through an opening.

(12) *Bend.* A bend (in this manual called a knot) is used to fasten two ropes together or to fasten a rope to a ring or loop.

(13) *Hitch.* A hitch is used to tie a rope around a timber, pipe, or post so that it will hold temporarily but can be readily undone.

b. Whipping Ends of Rope. The raw, cut end of a rope has a tendency to untwist, and should always be knotted or fastened in some manner to prevent this untwisting. Whipping (fig. 2–2) is one method of fastening the end of the rope to prevent untwisting. A rope is whipped by wrapping the end tightly with a small cord. This method is particularly satisfactory because there is very little increase in the size of the rope. The whipped end of a rope will still thread through blocks or other openings. Before cutting a rope, place two whippings on the rope 1 or 2 inches apart and make the cut between the whippings (fig. 2–2). This will prevent the cut ends from untwisting immediately after they are cut.

2–2. Knots at the End of Rope

a. Overhand Knot. The overhand knot (fig.

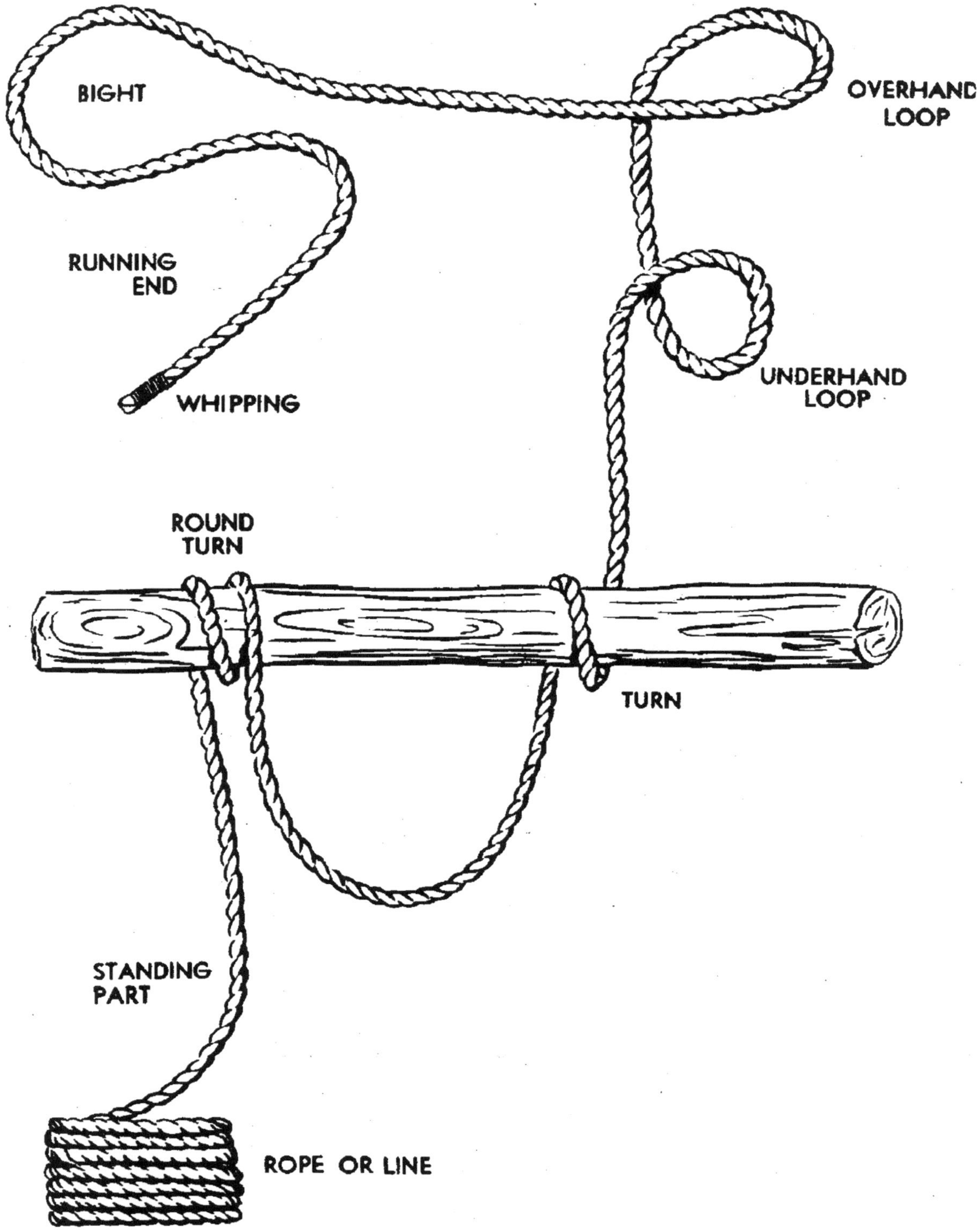

Figure 2–1. Elements of knots, bends, and hitches.

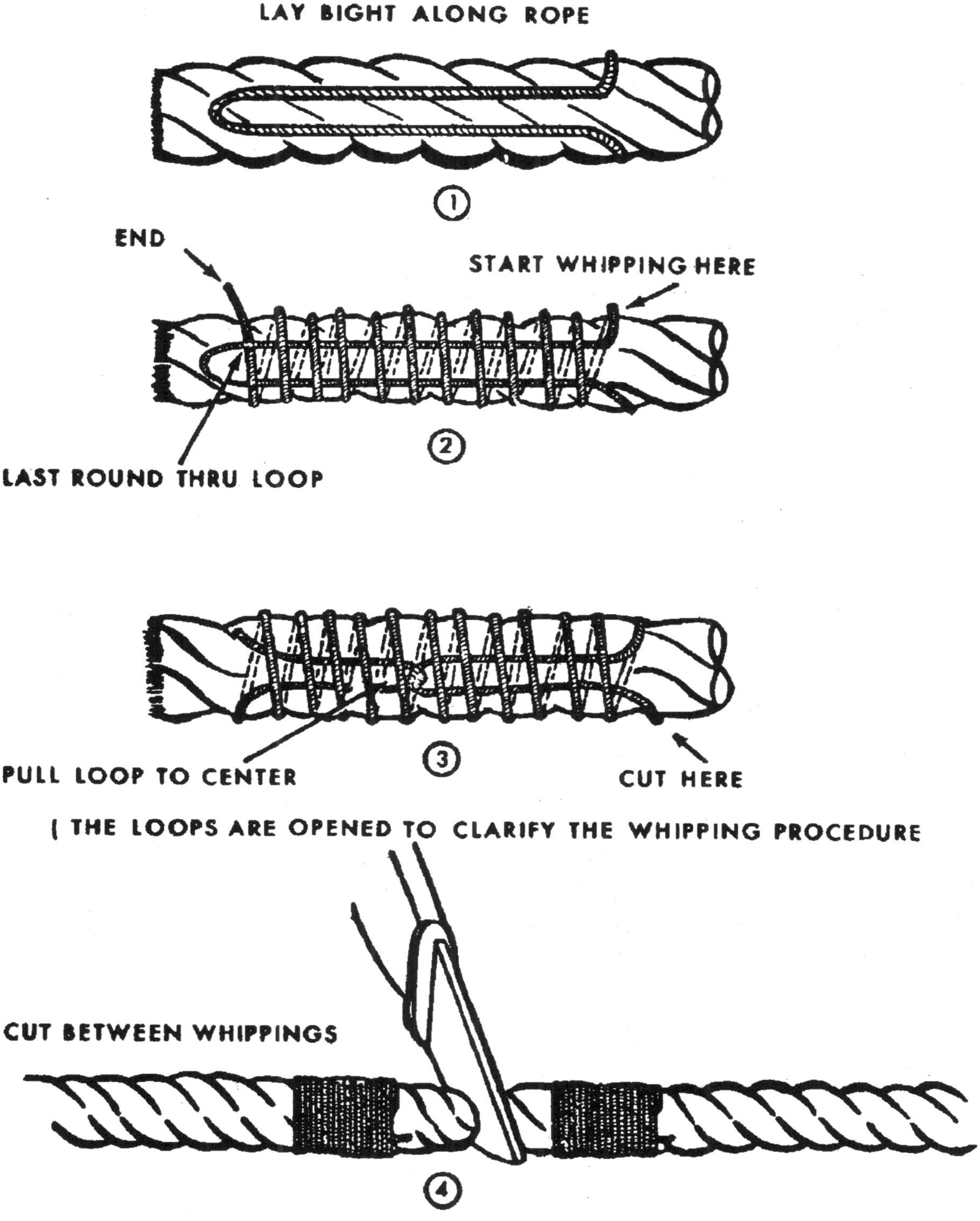

Figure 2 2. Whipping the end of a rope.

2–3) is the most commonly used and the simplest of all knots. An overhand knot may be used to prevent the end of a rope from untwisting, to form a knob at the end of a rope, or to serve as a part of another knot. When tied at the end or standing part of a rope, this knot prevents it from sliding through a block, hole, or another knot. It is also used to increase a person's grip on a rope. This knot reduces the strength of a straight rope by 55 percent.

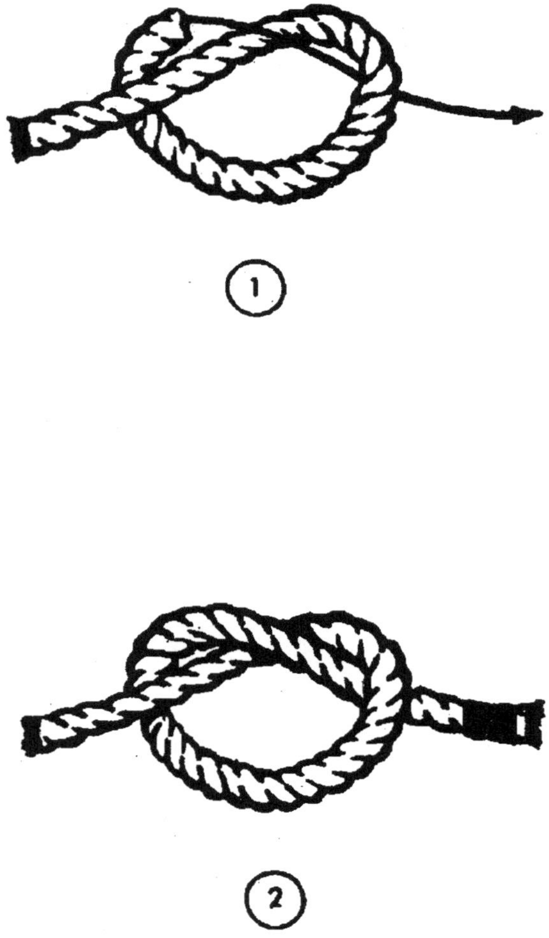

Figure 2–3. Overhand knot.

b. Figure Eight Knot. The figure eight knot (fig. 2–4) is used to form a larger knot at the end of a rope than would be formed by an overhand knot. A figure eight knot is used in the end of a rope to prevent the end from slipping through a fastening or loop in another rope, or from unreeving when reeved through blocks. It is easy to untie.

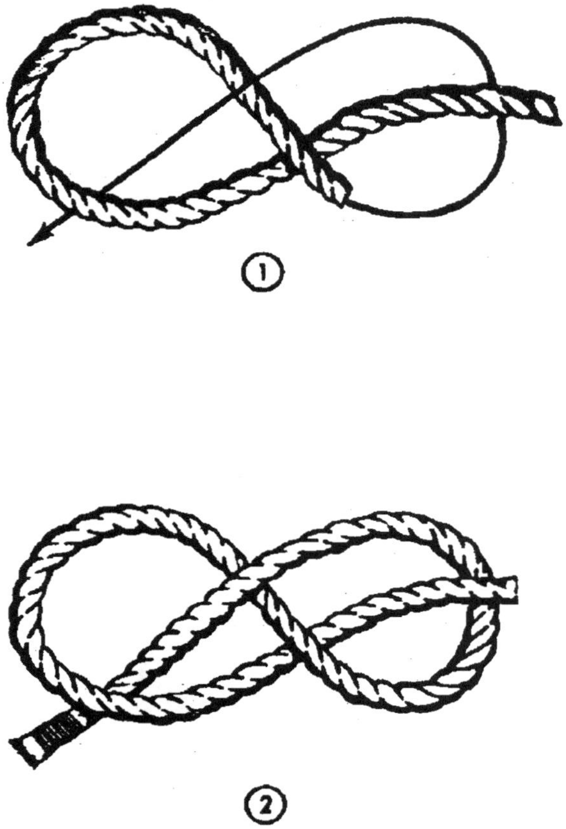

Figure 2–4. Figure eight knot.

c. Wall Knot. The wall knot (fig. 2–5) with crown is used to prevent the end of a rope from untwisting when an enlarged end is not objectionable. It also makes a desirable knot to prevent the end of the rope from slipping through small openings, as when rope handles are used on boxes. Either the crown or the wall knot may be used separately, to form semipermanent "stopper knots" tied with the end strands of a rope. The wall knot will prevent the rope from untwisting, but to make a neat round knob, it should be crowned (fig. 2–6). Notice that in the wall knot the ends come up through the bights, causing the strands to lead forward. In a crown knot the ends go down through the bights and point backward.

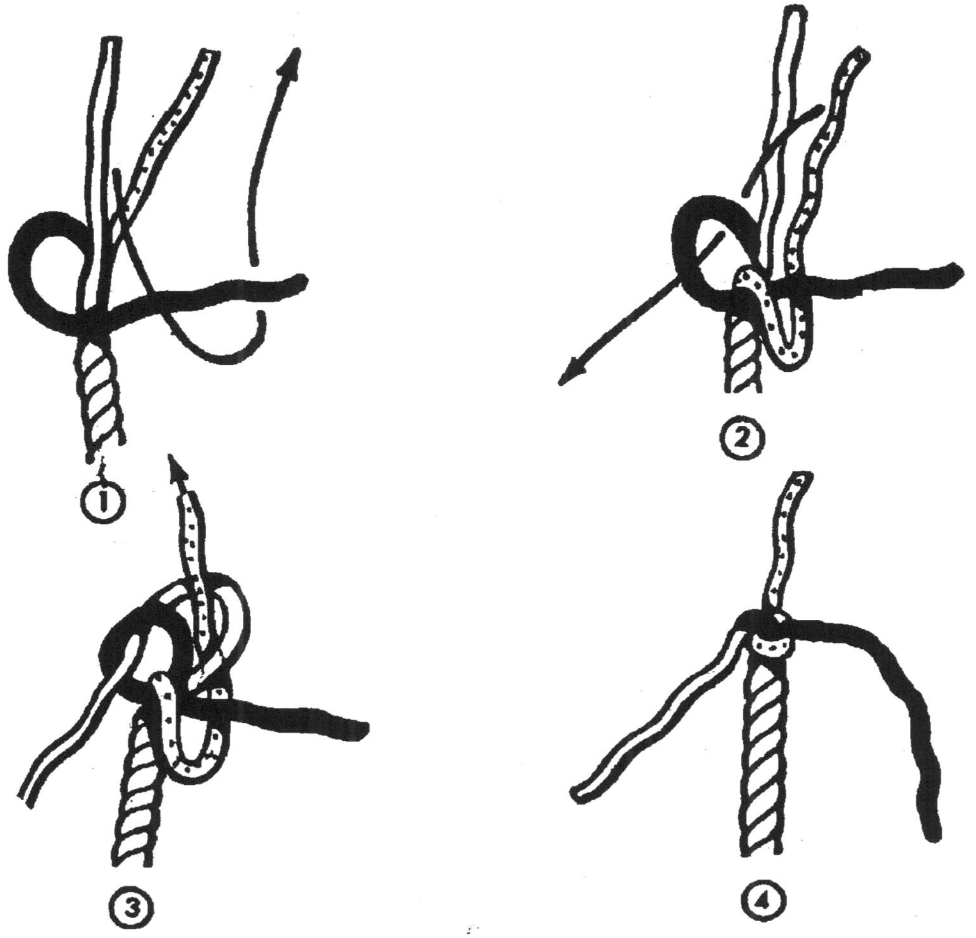

Figure 2–5. Wall knot.

2–3. Knots for Joining Two Ropes

a. Square Knot. The square knot (fig. 2–7) is used for tying two ropes of equal size together so they will not slip. Note that in the square knot the end and standing part of one rope comes out on the same side of the bight formed by the other rope. The square knot will not hold if the ropes are wet or if they are of different sizes. It tightens under strain but can be untied by grasping the ends of the two bights and pulling the knot apart.

Note. It makes no difference whether the first crossing is tied left-over-right or right-over-left, as long as the second crossing is tied opposite to the first crossing.

b. Single Sheet Bend. The use of a single sheet bend (fig. 2–8), sometimes called a weaver's knot, has two major uses (1) tying together two ropes of unequal size and (2) tying a rope to an eye. This knot will draw tight but will loosen or slip when the lines are slackened. The single sheet bend is stronger and more easily untied than the square knot.

c. Double Sheet Bend. The double sheet bend (fig. 2–9) has greater holding power than the single sheet bend for joining ropes of equal or unequal diameter, joining wet ropes, or tying a rope to an eye. It will not slip or draw tight under heavy loads. This knot is more secure than the single sheet bend when used in a spliced eye.

d. Carrick Bend. The carrick bend (fig. 2–10) is used for heavy loads and for joining large hawsers or heavy rope. It will not draw

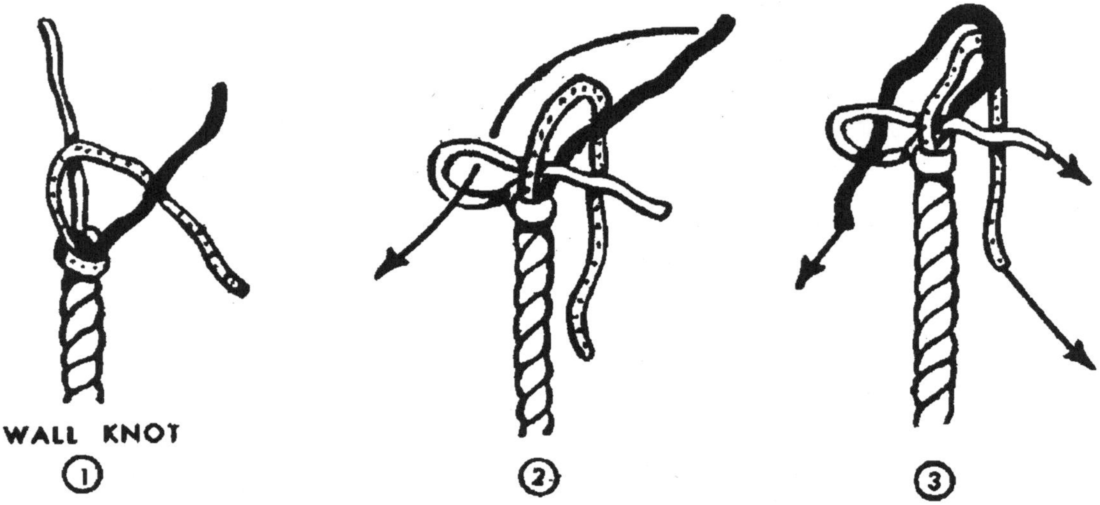

Figure 2–6. Crown on wall knot.

tight under a heavy load and is easily untied if the ends are seized to their own standing part.

2–4. Knots for Making Loops

a. Bowline. The bowline (fig. 2–11) is one of the most common knots and has a variety of uses, one of which is the lowering of men and material. It is the best knot for forming a single loop that will not tighten or slip under strain, and is easily untied if each running end is seized to its own standing part. The bowline forms a loop which may be of any length.

b. Double Bowline. The double bowline (fig. 2–12) forms 3 nonslipping loops. This knot can be used for slinging a man. As he sits in the slings, one loop is used to support his back and the remaining two loops support his legs; a notched board passed through the two loops makes a comfortable seat known as a boatswains chair. This chair is discussed in the scaffolding section of this manual (chap 6).

c. Running Bowline. The running bowline (fig. 2–13) forms a strong running loop. It is a convenient form of running an eye. The running bowline provides a sling of the choker type at the end of a single line. It is used when a handline is to be tied around an object at a point that cannot be safely reached, such as the end of a limb.

d. Bowline on a Bight. This knot (fig. 2–14) forms two nonslipping loops. The bowline on a bight can be used for the same purpose as a boatswain's chair. It does not leave both hands free, but its twin nonslipping loops form a comfortable seat. It is used when a greater strength than that given by a single bowline is necessary, when it is desirable to form a loop at some point in a rope other than at the end or when the end of a rope is not accessible. The bowline on a bight is easily untied, and can be tied at the end of a rope by doubling the rope for a short section.

e. Spanish Bowline. A Spanish bowline (fig. 2–15) can be tied at any point in a rope, either at a place where the line is double or at an end which has been doubled back. The Spanish bowline is used in rescue work or to give a twofold grip for lifting a pipe or other round objects in a sling.

f. French Bowline. The French bowline (fig. 2–16) is sometimes used as a sling for lifting injured men. When used for this purpose, one loop is used as a seat and the other loop is put around the body under the arms. The weight of the injured man keeps the two loops tight so that he cannot fall out. It is particularly useful as a sling for an insensible man. The French bowline may also be used where a man is working alone and needs both hands free. The two loops of this knot can be adjusted to the size required.

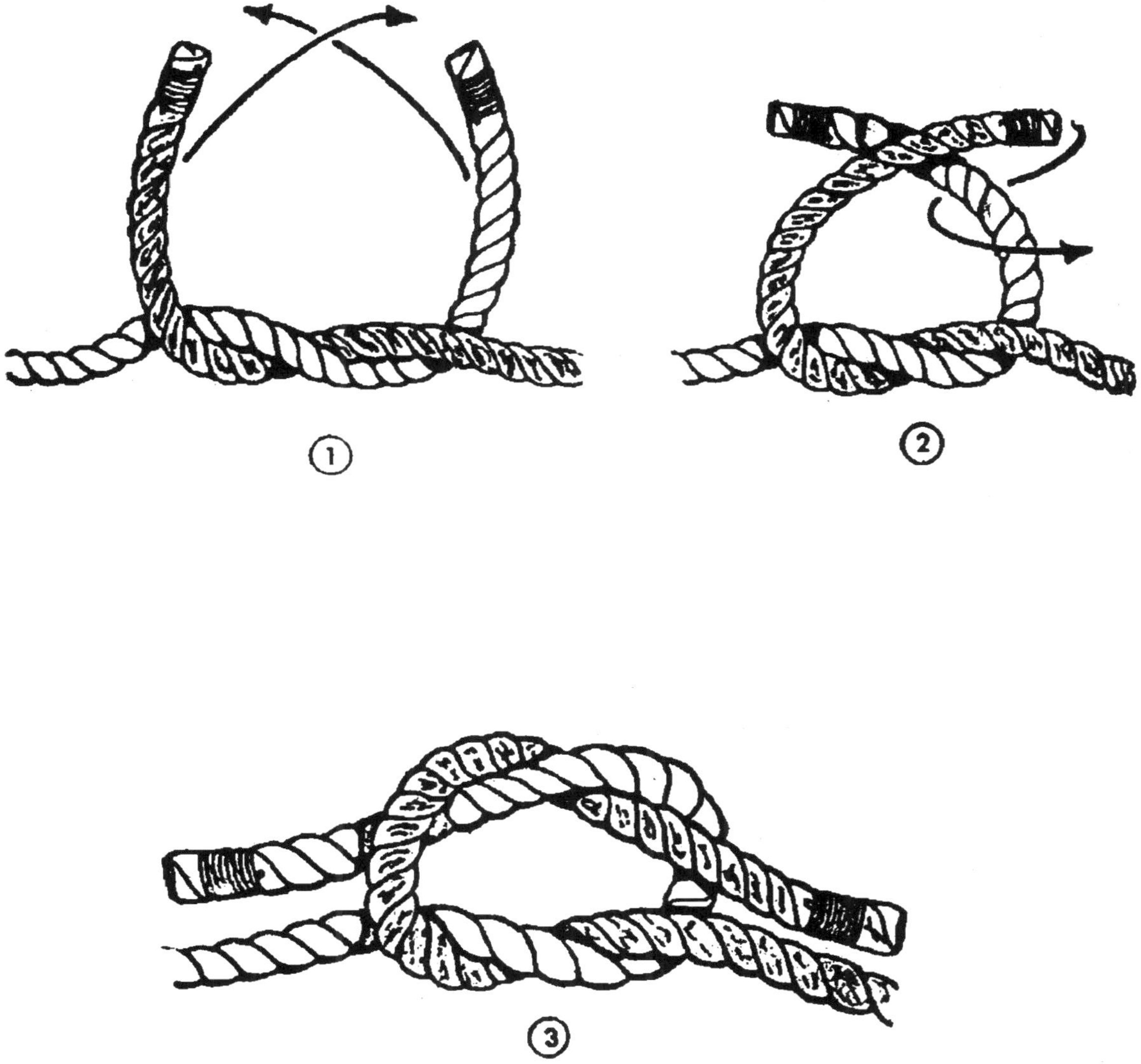

Figure 2–7. Square knot.

g. Speir Knot. A Speir knot (fig. 2–17) is used when a fixed loop, a nonslip knot, and a quick release are required. It can be tied quickly and released by a pull on the running end.

h. Catspaw. A catspaw (fig. 2–18) can be used for fastening an endless sling to a hook, or it can be made at the end of a rope for fastening the rope to a hook. It is easily tied or untied. This knot, which is really a form of hitch, is a more satisfactory way of attaching a rope to a hook than the blackwall hitch (para 2–5*k*). It will not slip off and need not be kept taut to make it hold.

i. Figure Eight With an Extra Turn. A figure eight with an extra turn (fig. 2–19) can be used to tighten a rope. This knot is especially well suited for tightening a one-rope bridge across a small stream. It is easily tied and untied.

2–5. Hitches

a. Half Hitch. The half hitch (A, fig. 2–20)

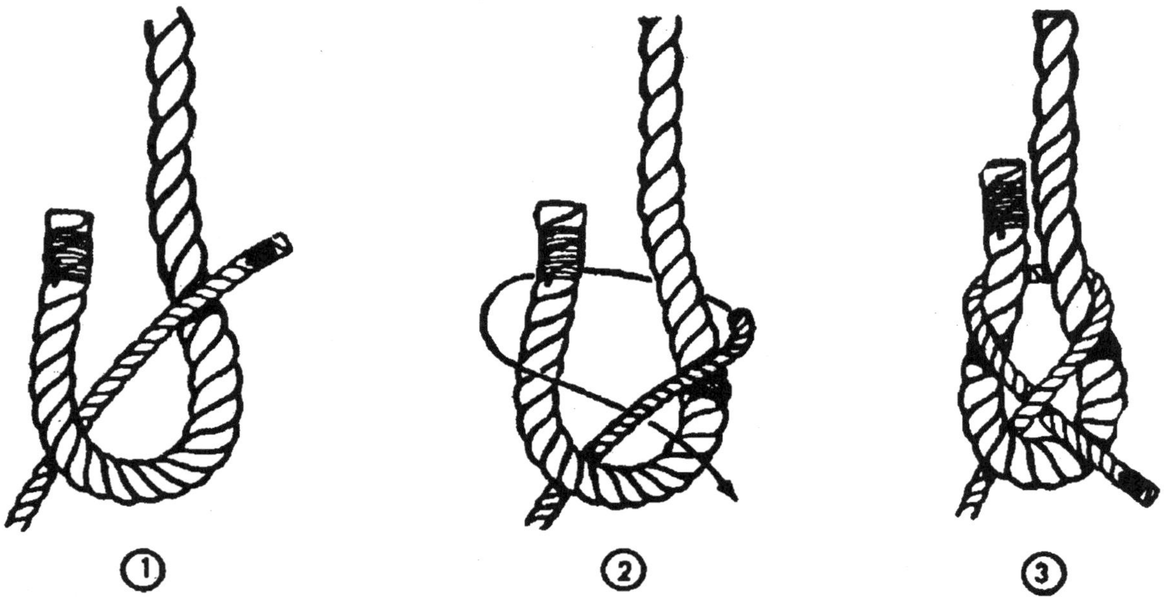

Figure 2–8. Single sheet bend.

Figure 2–9. Double sheet bend.

is used to tie a rope to a timber or to a larger rope. It will hold against a steady pull on the standing part of the rope, but is not a secure hitch. It is frequently used for securing the free end of a rope, and is an aid and the foundation of many knots. For example, it is the start of a timber hitch and a part of the fisherman's knot. It also makes the rolling hitch more secure.

b. Two Half Hitches. Two half hitches (B, fig. 2–20) are especially useful for securing the running end of a rope to the standing part. If the two hitches are slid together along the standing part to form a single knot, the knot becomes a clove hitch.

c. Round Turn and Two Half Hitches. Another hitch used for fastening a rope to a pole, timber, or spar is the round turn and two half

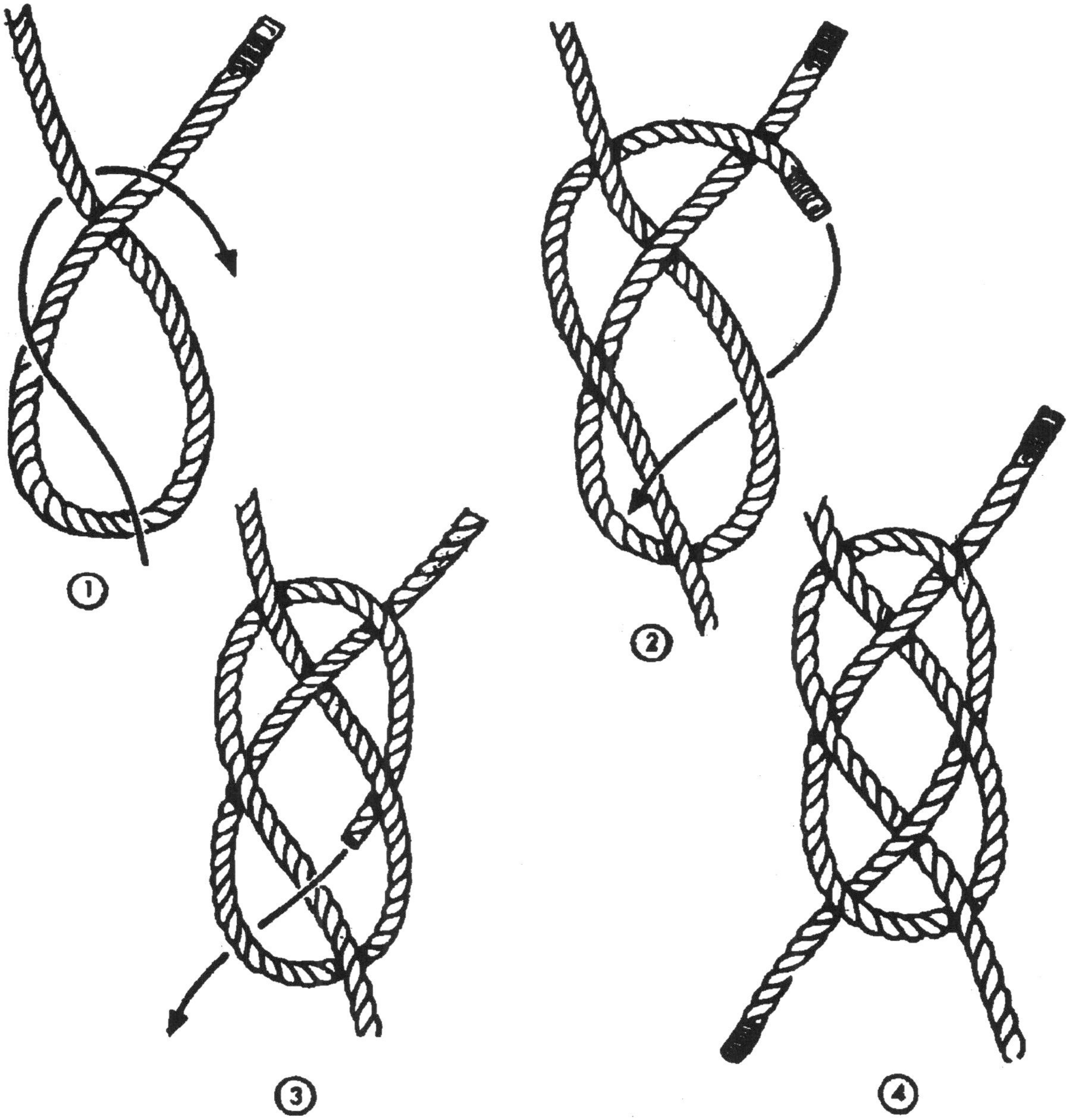

Figure 2–10. Carrick bend.

hitches (fig. 2–21). For greater security, the running end of the rope should be seized to the standing part. This hitch does not jam.

d. Timber Hitch. The timber hitch (fig. 2–22) is used for moving heavy timber or poles. This hitch is excellent for securing a piece of lumber or similar objects. The pressure of the coils, one over the other, holds the timber securely; the more tension applied, the tighter the hitch becomes about the timber. It will not slip, but will readily loosen when strain is relieved.

e. Timber Hitch and Half Hitch. A timber hitch and half hitch (fig. 2–23) are combined

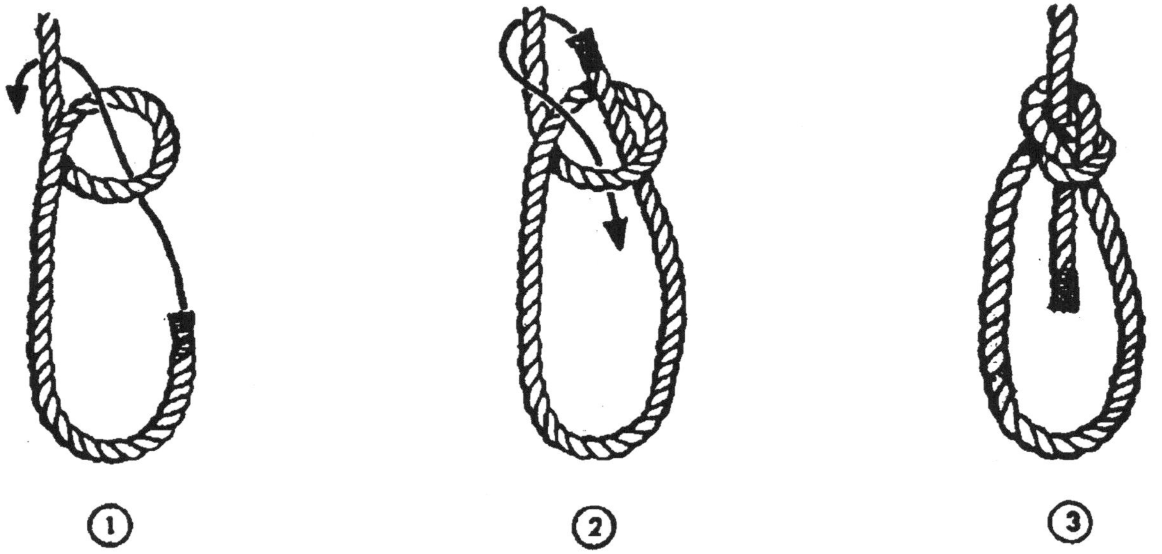

Figure 2–11. Bowline.

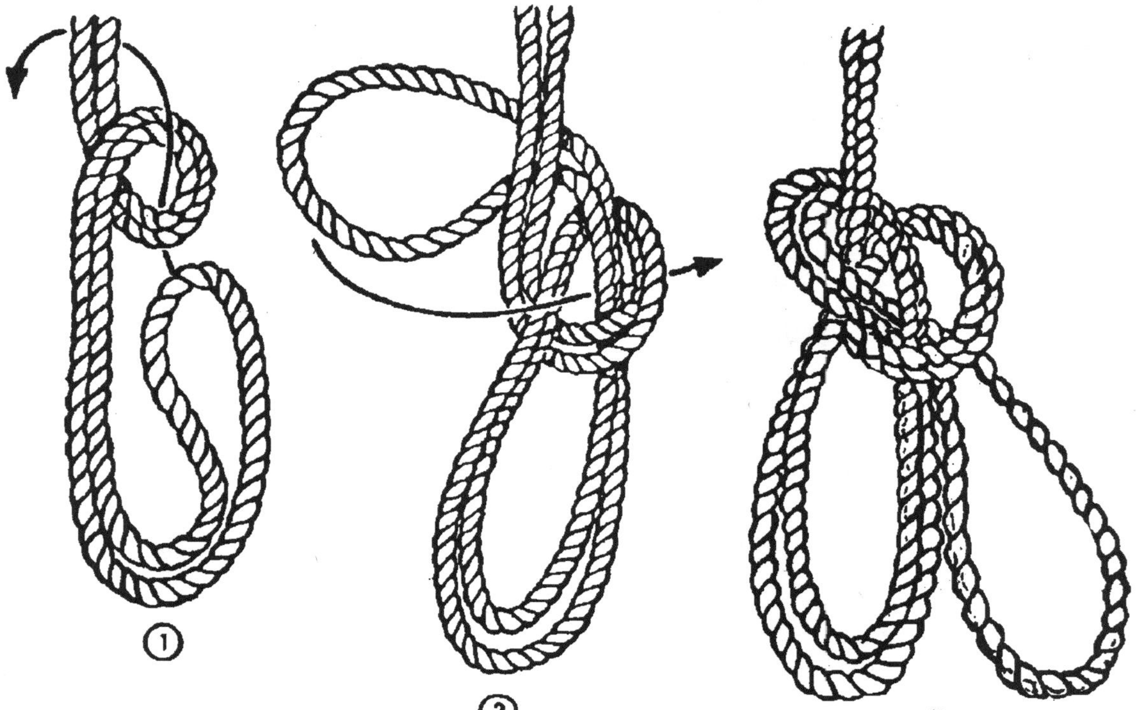

Figure 2–12. Double bowline.

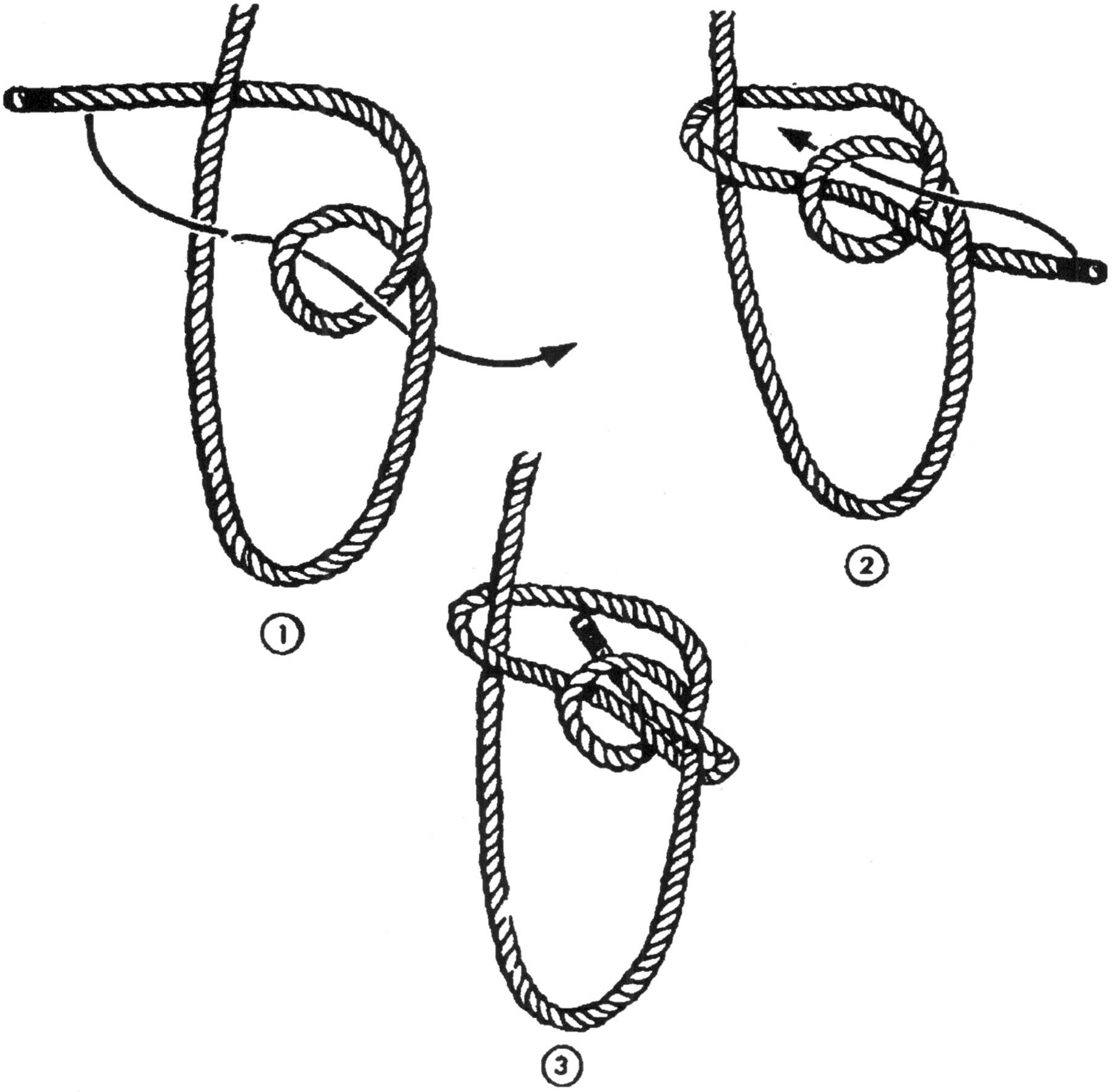

Figure 2–13. Running bowline.

to hold heavy timber or poles when they are being lifted or dragged. A timber hitch used alone may become untied when the rope is slack or a sudden strain is put on it.

f. Clove Hitch. The clove hitch (fig. 2–24) is one of the most widely used knots. It is used to fasten a rope to a timber, pipe, or post. It is also used for making other knots. This knot puts very little strain on the fibers when the rope is put around an object in one continuous direction. The clove hitch can be tied at any point in a rope. If there isn't constant tension on the rope, another loop (round of the rope around the object and under the center of the clove hitch will permit a tightening and slackening motion of the rope.

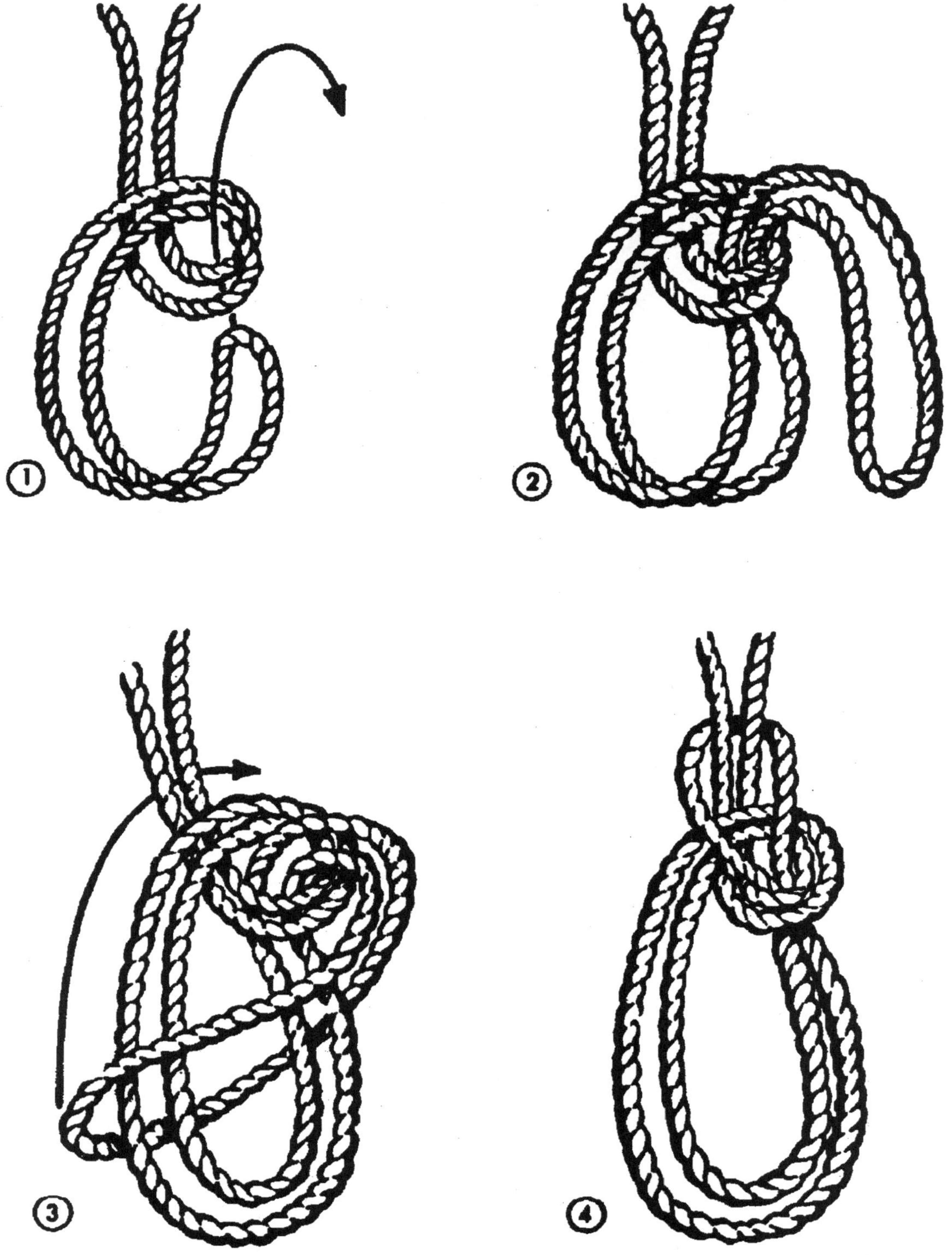

Figure 2-14. Bowline on a bight.

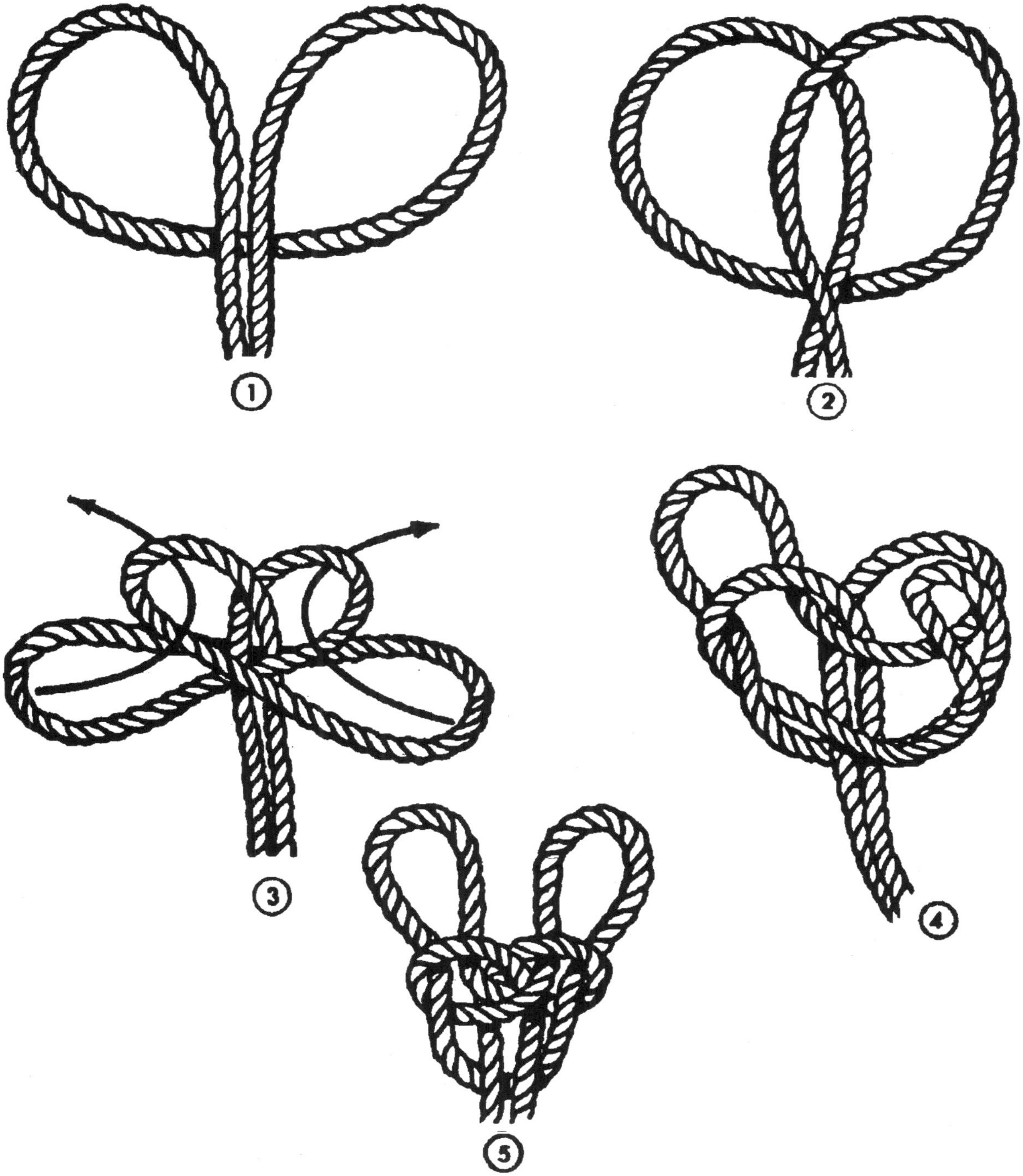

Figure 2-15. Spanish bowline.

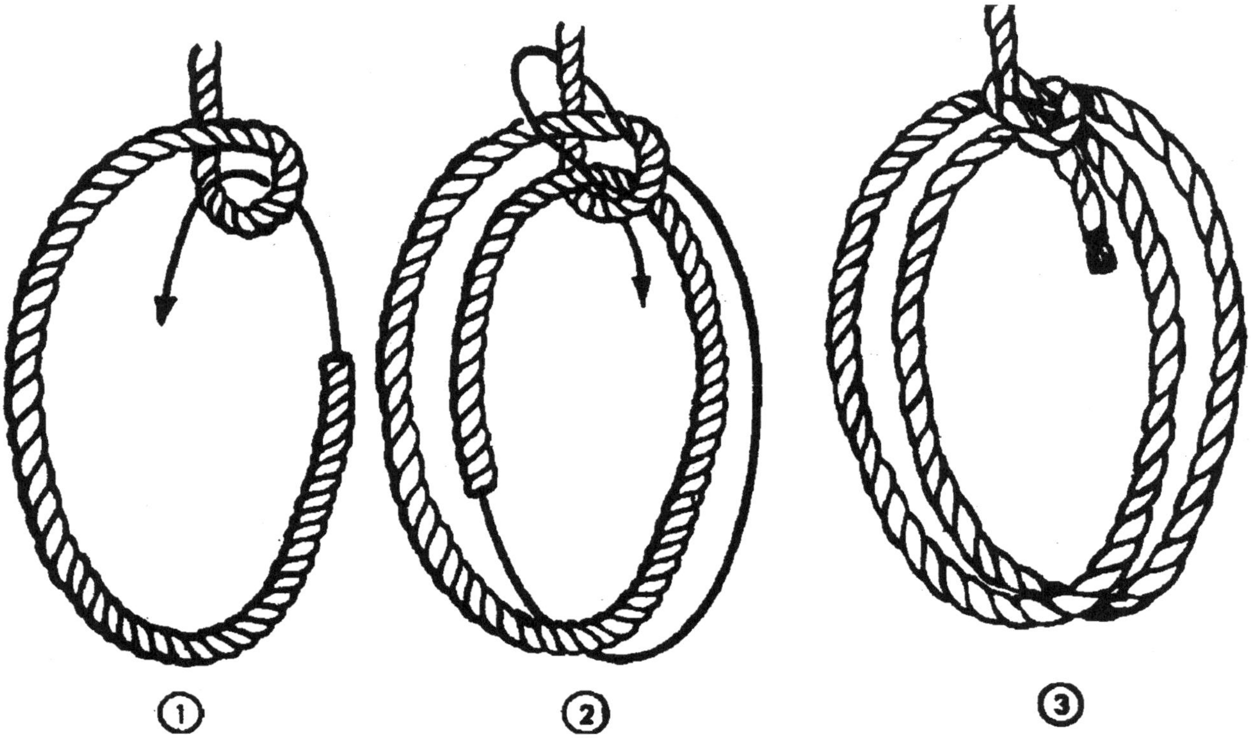

Figure 2–16. French bowline.

g. Rolling Hitch. The rolling hitch (fig. 2–25) is used to secure a rope to another rope, or fasten it to a pole or pipe so that the rope will not slip. This knot grips tightly, but is easily moved along a rope or pole when strain is relieved.

h. Telegraph Hitch. The telegraph hitch (fig. 2–26) is a very useful and secure hitch which is used to hoist or haul posts and poles. It is easy to tie and untie, and will not slip.

i. Mooring Hitch. The mooring hitch (fig. 2–27), also called rolling or magnus hitch, is used to fasten a rope around a mooring post or to attach a rope at a right angle to a post. This hitch grips tightly and is easily removed.

j. Scaffold Hitch. The scaffold hitch (fig. 2–28) is used to support the end of a scaffold plank with a single rope. It prevents the plank from tilting.

k. Blackwall Hitch. The blackwall hitch (fig. 2–29) is used for fastening a rope to a hook. It is generally used to attach a rope temporarily to a hook or similar object in derrick work. This hitch holds only when subjected to a constant strain or when used in the middle of a rope with both ends secured. Human life and breakable equipment should never be entrusted to the blackwall hitch.

l. Harness Hitch. The harness hitch (fig. 2–30) forms a nonslipping loop in a rope. It is often employed by putting an arm through the loop, then placing the loop on the shoulder and pulling the object attached to the rope. The hitch is tied only in the middle of a rope. It will slip if only one end of the rope is pulled.

m. Girth Hitch. The girth hitch (fig. 2–31) is used in tying suspender ropes to hand ropes in the construction of expedient foot bridges. It is a simple and convenient hitch for many other uses of ropes and cords.

n. Sheepshank. A sheepshank (fig. 2–32) is a method of shortening a rope, but it also may be used to take the load off a weak spot in the rope. It is only a temporary knot unless the

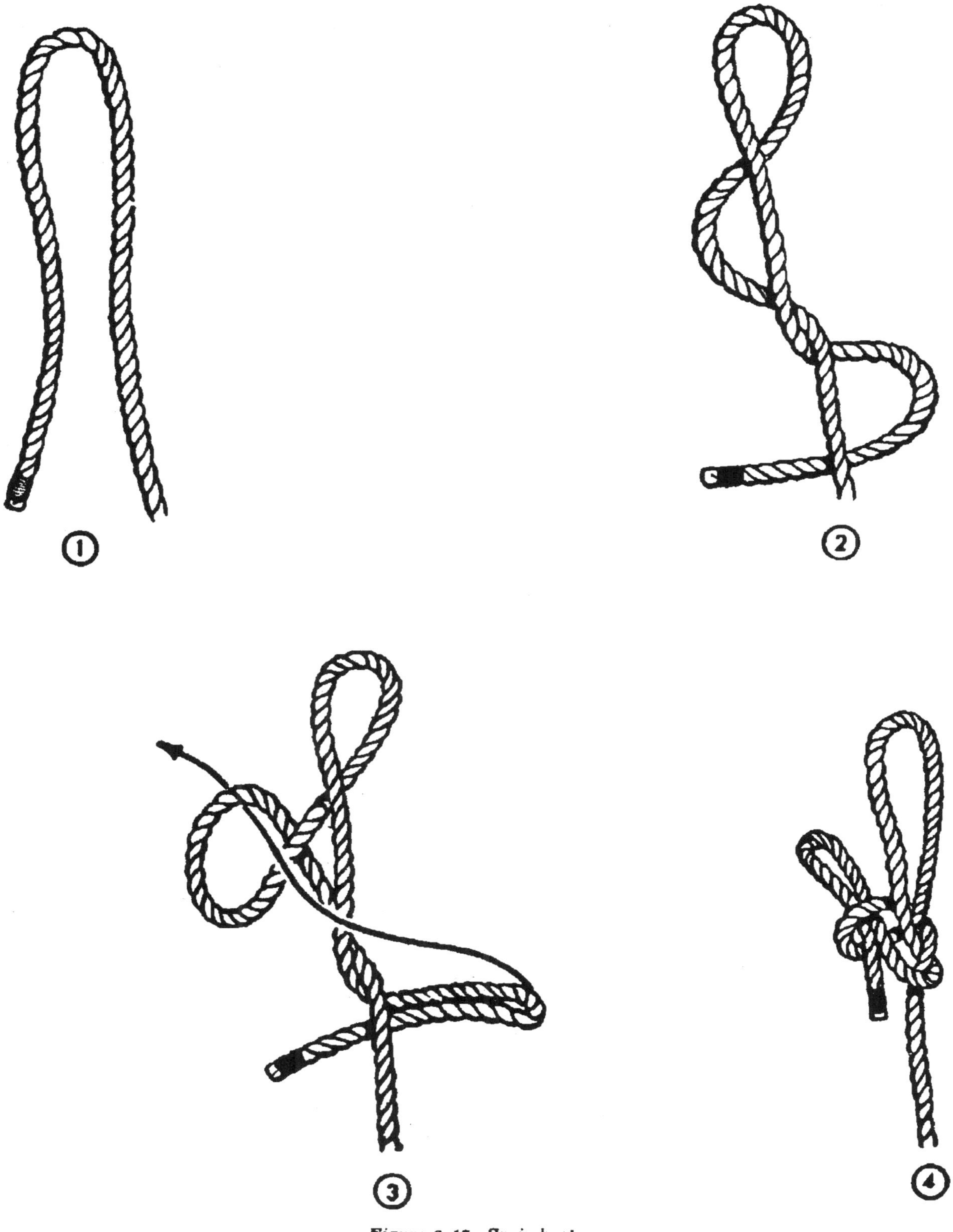

Figure 2–17. Speir knot.

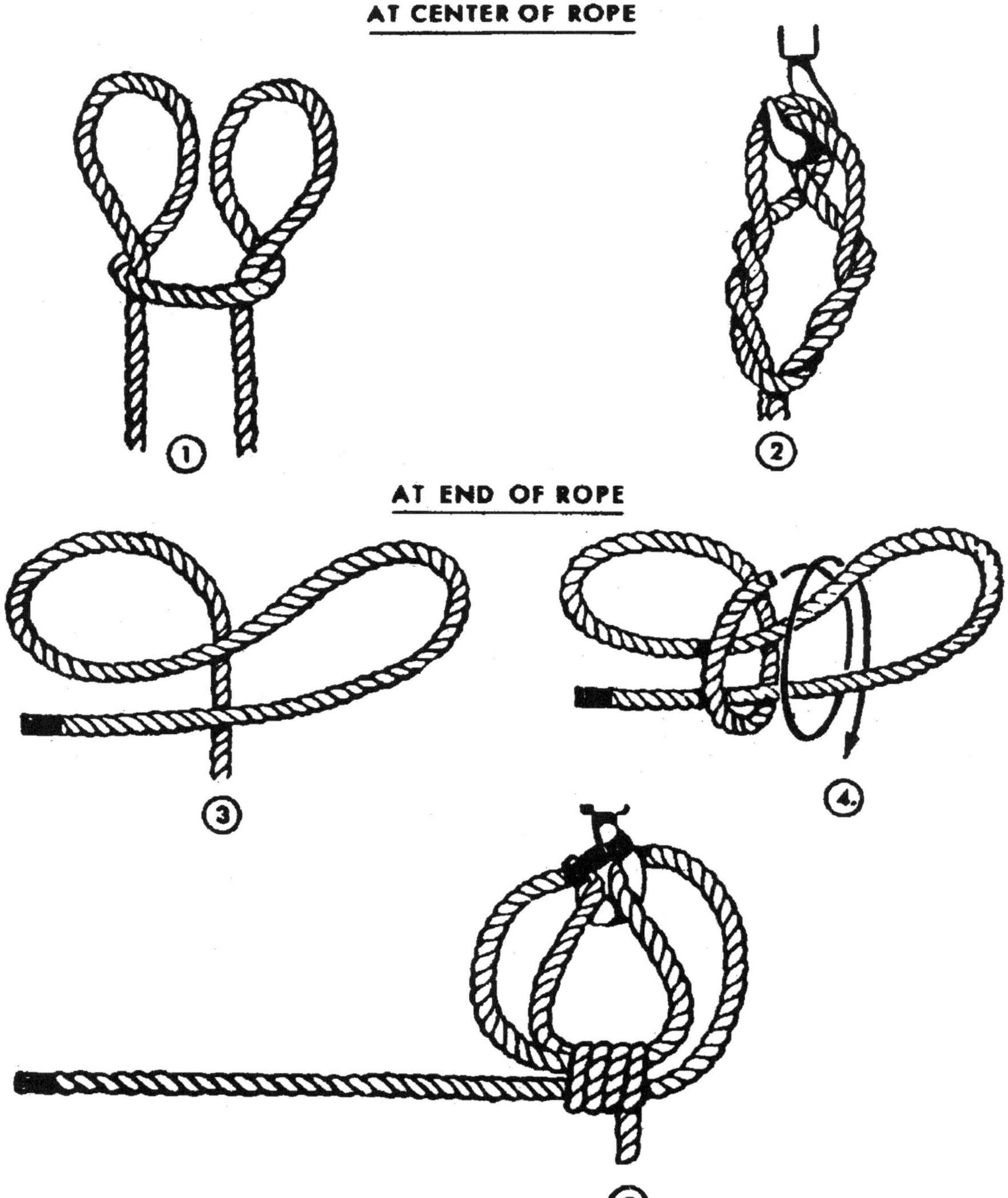

Figure 2-18. Catspaw.

Figure 2 19. Figure eight with extra turn.

eyes are fastened to the standing part on each end.

o. Fisherman's Bend. The fisherman's bend (fig. 2–33) is an excellent knot for attaching a rope to a light anchor, a ring, or a rectangular piece of stone. It can be used to fasten a rope or cable to a, ring or post or where there will be slackening and tighening motion in the rope.

2–6. Knots for Tightening a Rope

a. Butterfly Knot. The butterfly knot (fig. 2–34) is used to pull taut a high line, handline, tread rope for foot bridges, or similar installations. Use of this knot will provide the capability to tighten a fixed rope when mechanical means are not available. (The harness hitch (fig. 2–30) can also be used for this purpose.) The butterfly knot will not jam if a stick is placed between the two upper loops.

b. Baker Bowline. The baker bowline (fig. 2–35) may be used for the same purpose as the butterfly knot (fig. 2–34) and for lashing cargo. When used to lash cargo, secure one end with two half hitches, pass the rope over the cargo and tie a baker bowline, then secure the lashing with a slippery half hitch. To release the rope, simply pull on the running end. The baker bowline has the advantage of being easy

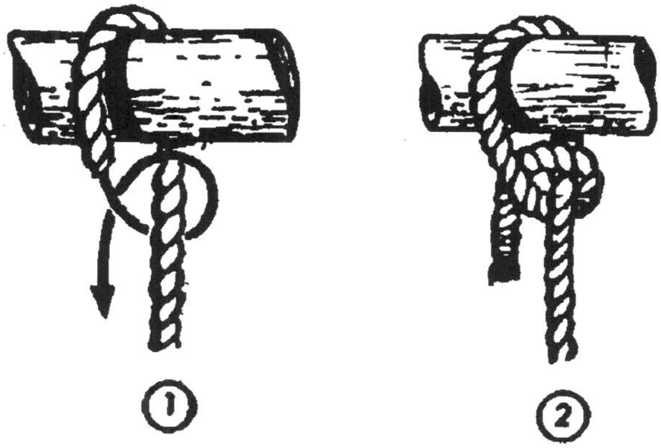

<u>Half hitch.</u>

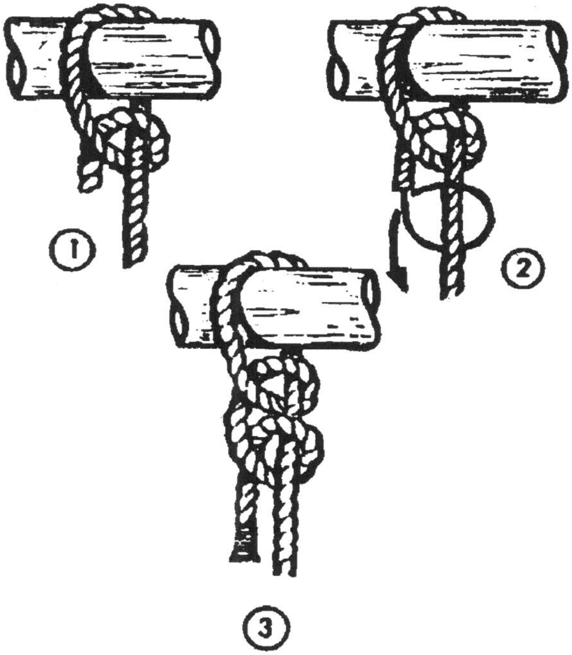

Figure 2–20. Half hitches.

to tie, can be adjusted without losing control, and can be released quickly.

2–7. Lashings

a. Square Lashing. The square lashing (fig. 2–36) is used to lash two spars together at right angles to each other. To tie a square lashing, begin with a clove hitch on one spar and

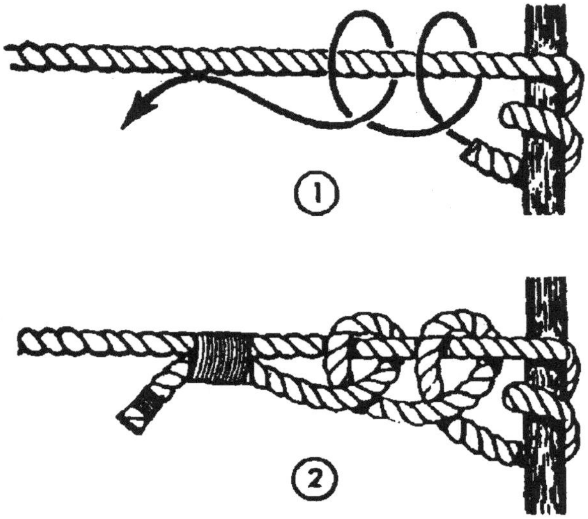

Figure 2 21. Round turn and two half hitches.

Figure 2–22. Timber hitch.

Figure 2–23. Timber hitch and half hitch.

make a minimum of 4 complete turns around both members. Continue with two frapping turns between the vertical and the horizontal spar to tighten the lashing. Tie off the running end to the opposite spar from which you started with another clove hitch to finish the square lashing.

b. Shears Lashing. The shears lashing (fig. 2–37) is used to lash 2 spars together at one end to form an expedient device called a shears. This is done by laying 2 spars side by side, spaced apprximately ⅓ the diameter of a spar apart, with the butt ends together. The shears lashing is started a short distance in from the top of one of the spars by tying the end of the rope to it with a clove hitch. Then 8 tight turns are made around both spars above the clove hitch. The lashing is tightened with a minimum of 2 frapping turns around the 8 turns. The shears lashing is finished by tying the end of the rope to the opposite spar from which you started with another clove hitch.

c. Block Lashing. Block lashing (fig. 2–38) is used to tie a tackle block to a spar. First, 3 right turns of the rope are made around the spar where the tackle block is to be attached. The next 2 turns of the rope are passed through the mouth of the hook or shackle of the tackle block and drawn tightly. Then 3 additional taut turns of the rope are put around the spar above the hook or shackle. The block lashing is completed by tying the 2 ends of the rope together with a square knot. When a sling is supported by a block lashing, the sling is passed through the center 4 turns.

2–8. Knots for Wire Rope

Under special circumstances when wire rope fittings are not available and it is necessary to fasten wire rope by some other manner, certain knots can be used. In all knots made with wire rope, the running end of the rope should be fastened to the standing part after the knot is tied. When wire rope clips are available they

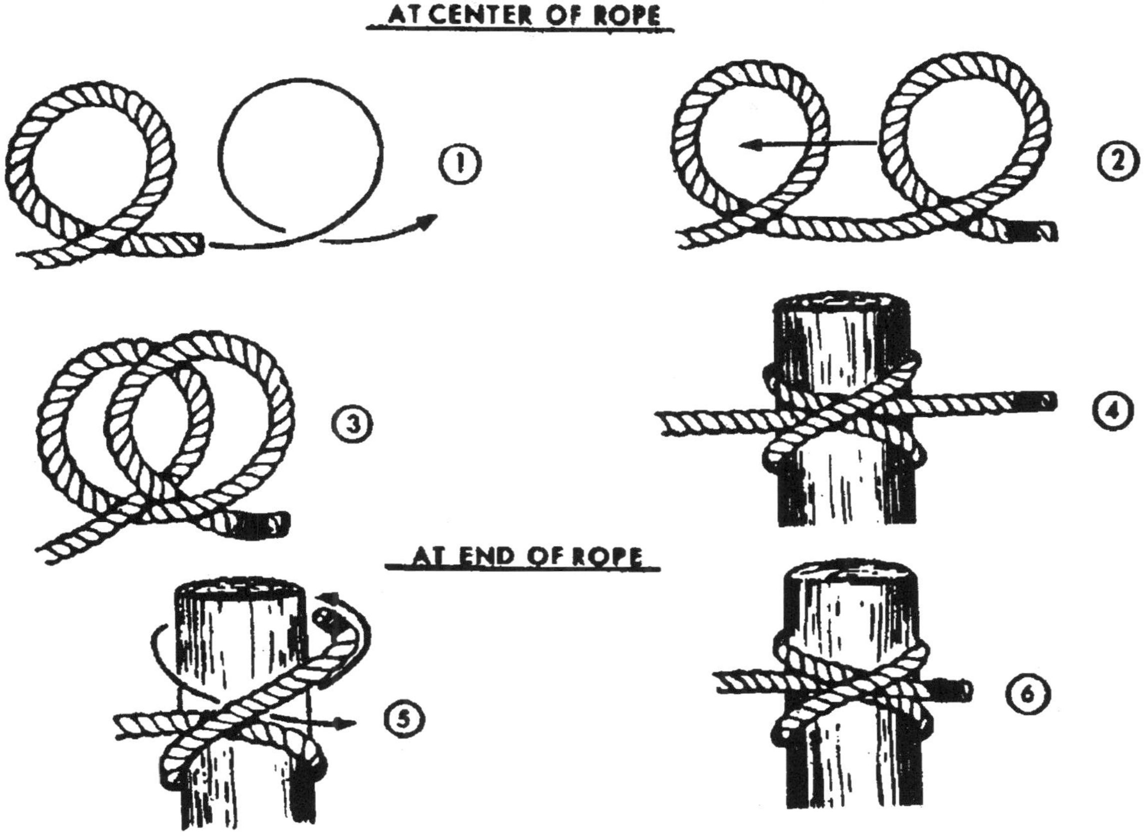

Figure 2–24. Clove hitch.

should be used for fastening the running end. If clips are not available, wire or strand of cordage may be used. All knots in wire rope should be checked periodically for wear or signs of breakage. If there is any reason to believe that the knot has been subjected to excessive wear, a short length of the end of the rope, including the knot, should be cut off and a new knot should be tied. The fisherman's bend, clove hitch, and carrick bend can be used for fastening wire rope.

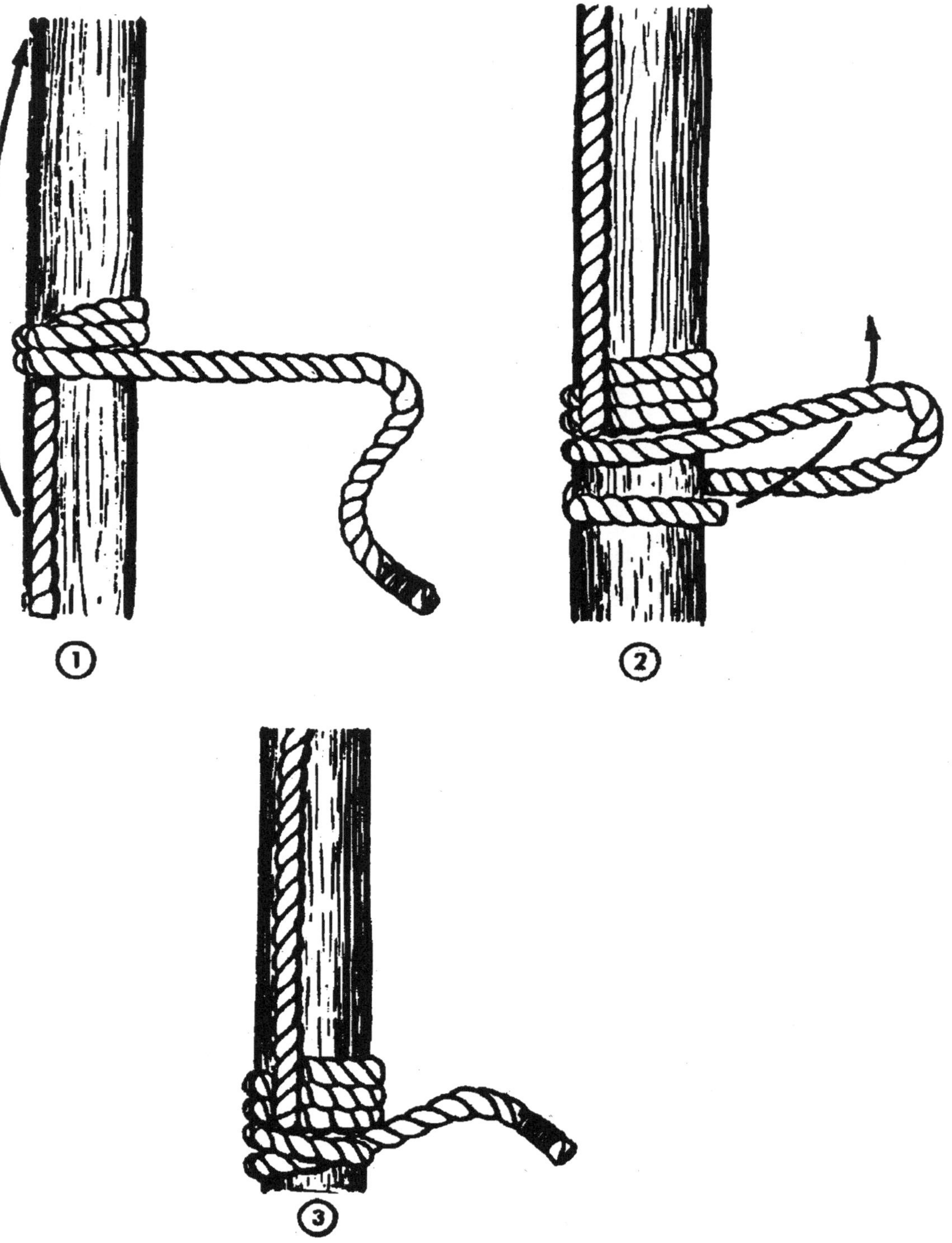

Figure 2–25. Rolling hitch.

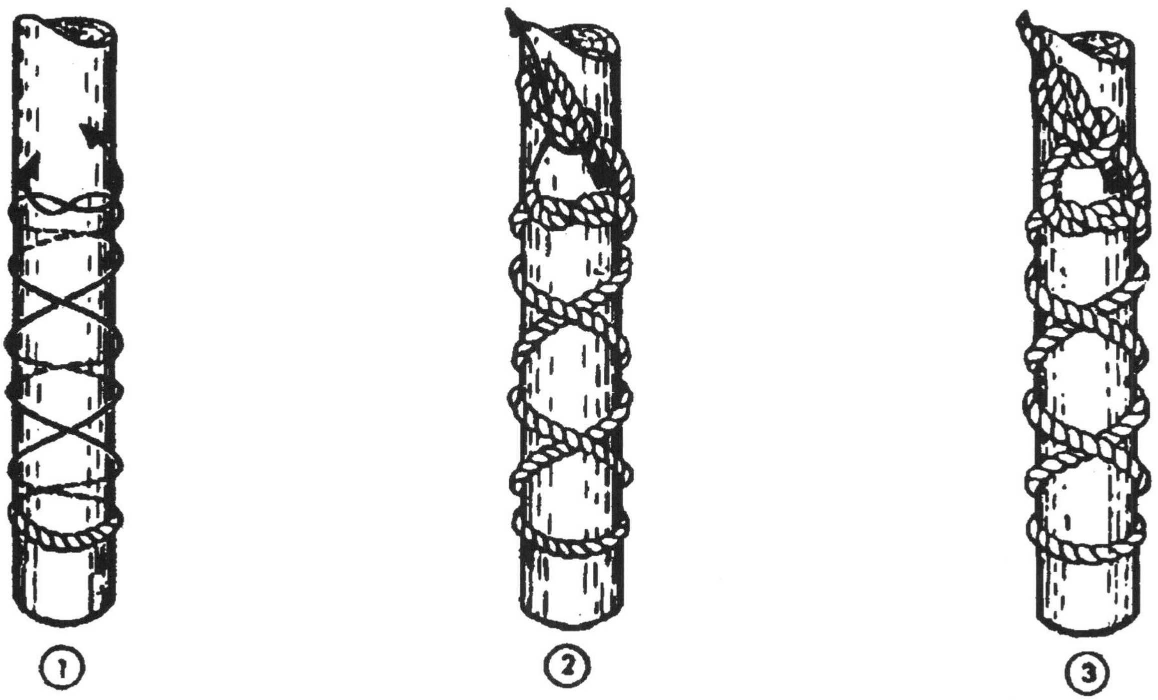

Figure 2–26. Telegraph kitch.

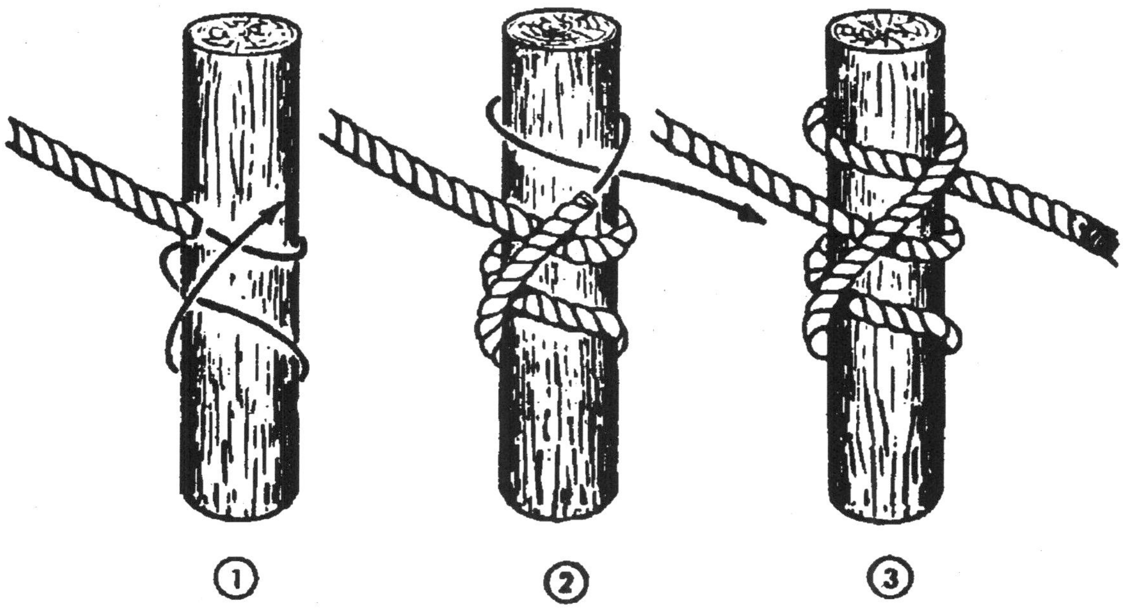

Figure 2–27. Mooring kitch.

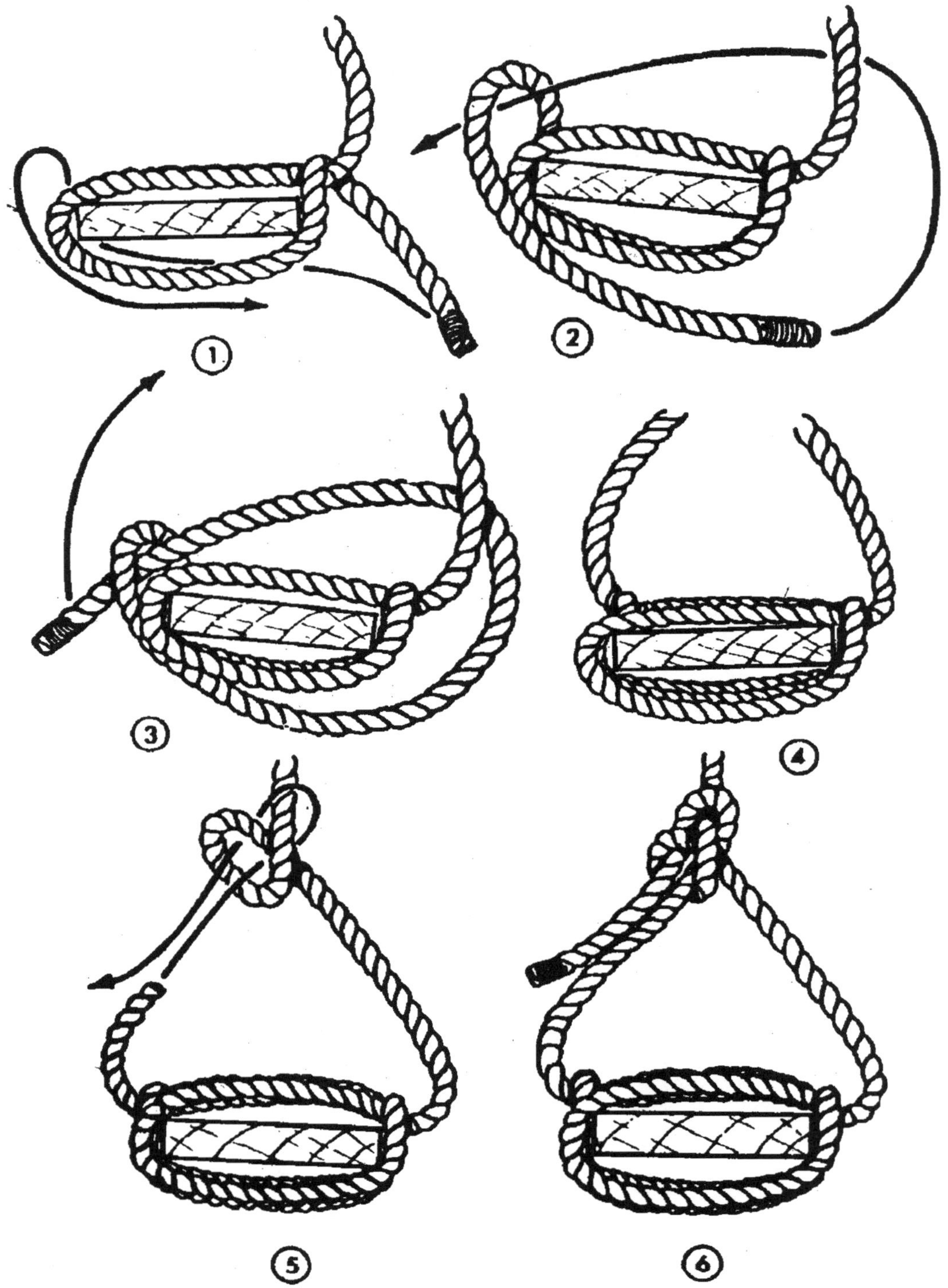

Figure 2 28. Scaffold hitch.

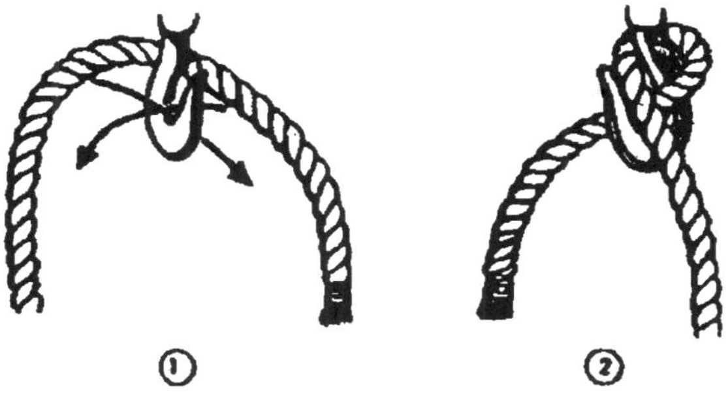

Figure 2–29. Blackwall hitch.

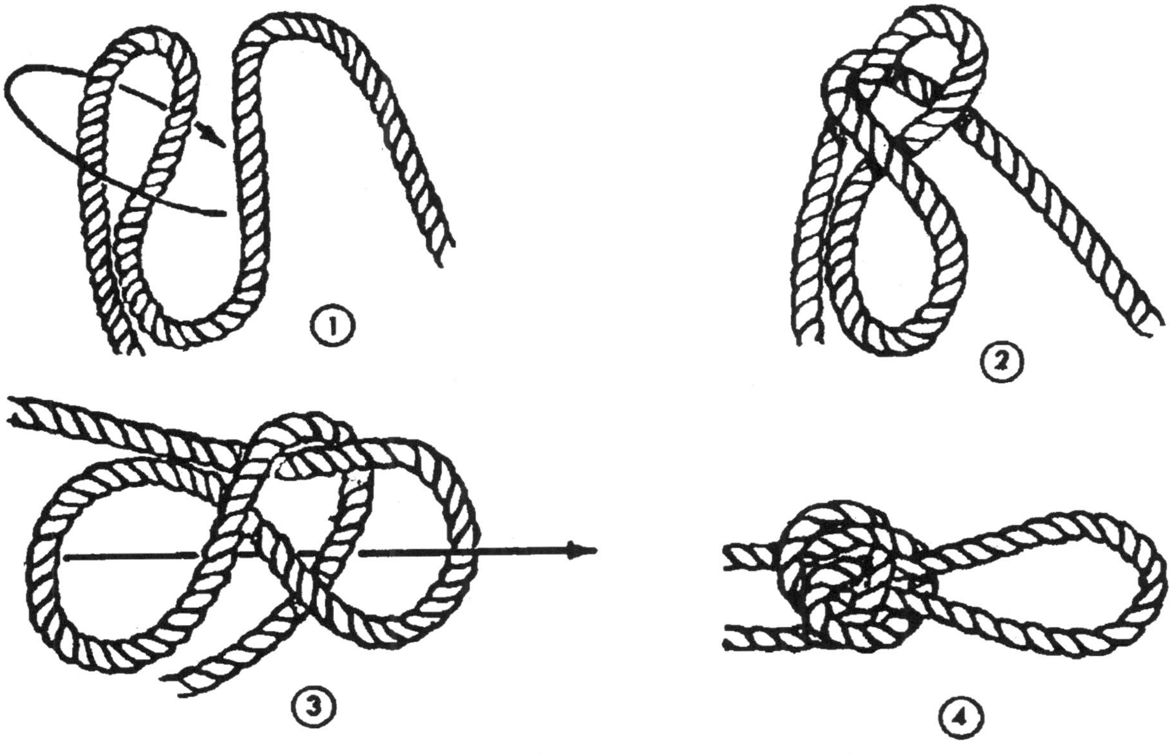

Figure 2–30. Harness hitch.

Figure 2–31. Girth hitch.

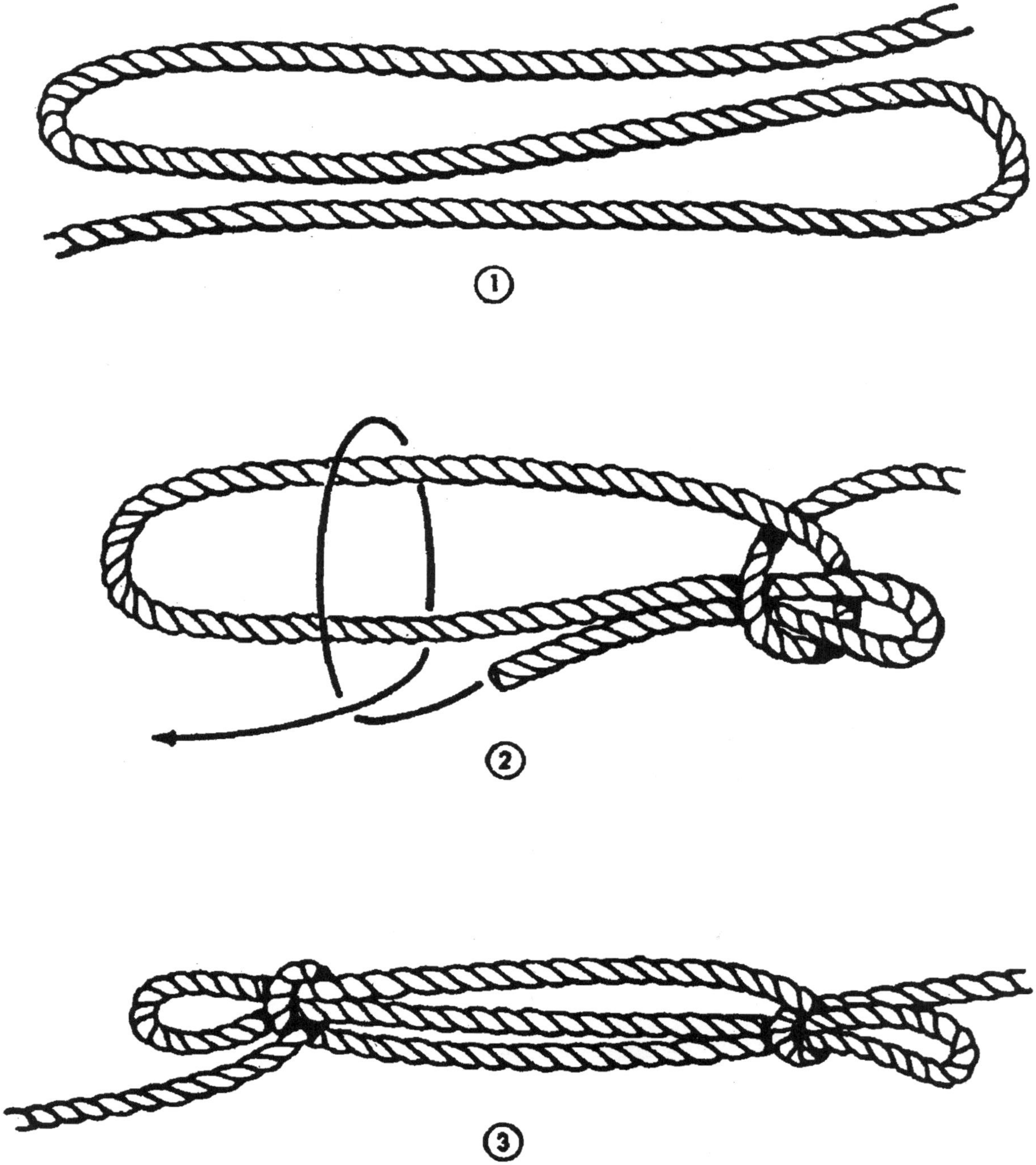

Figure 2–32. Sheepshank.

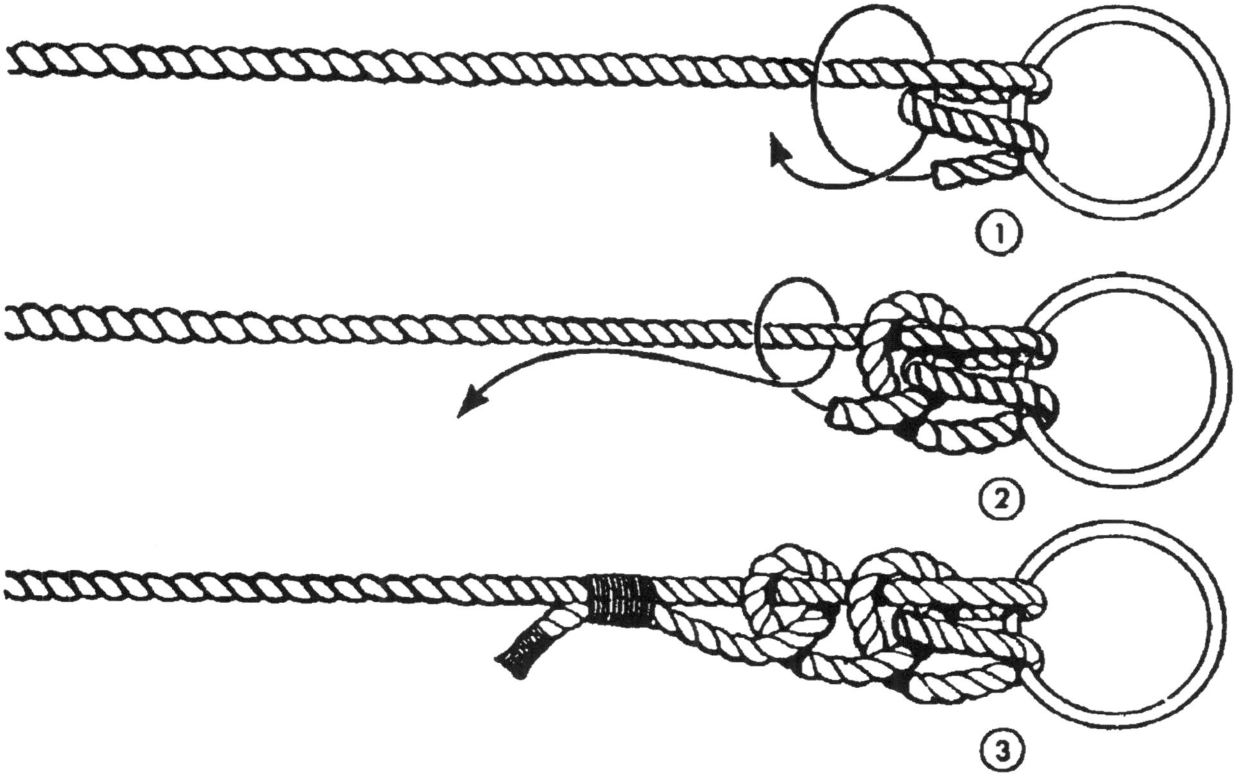

Figure 2–33. Fishermen's bend.

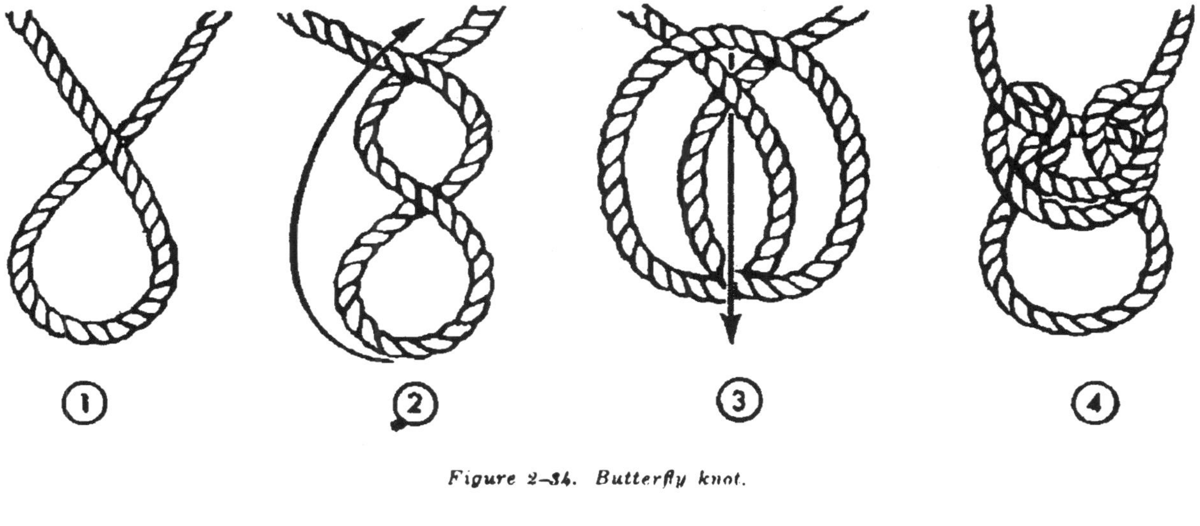

Figure 2–34. Butterfly knot.

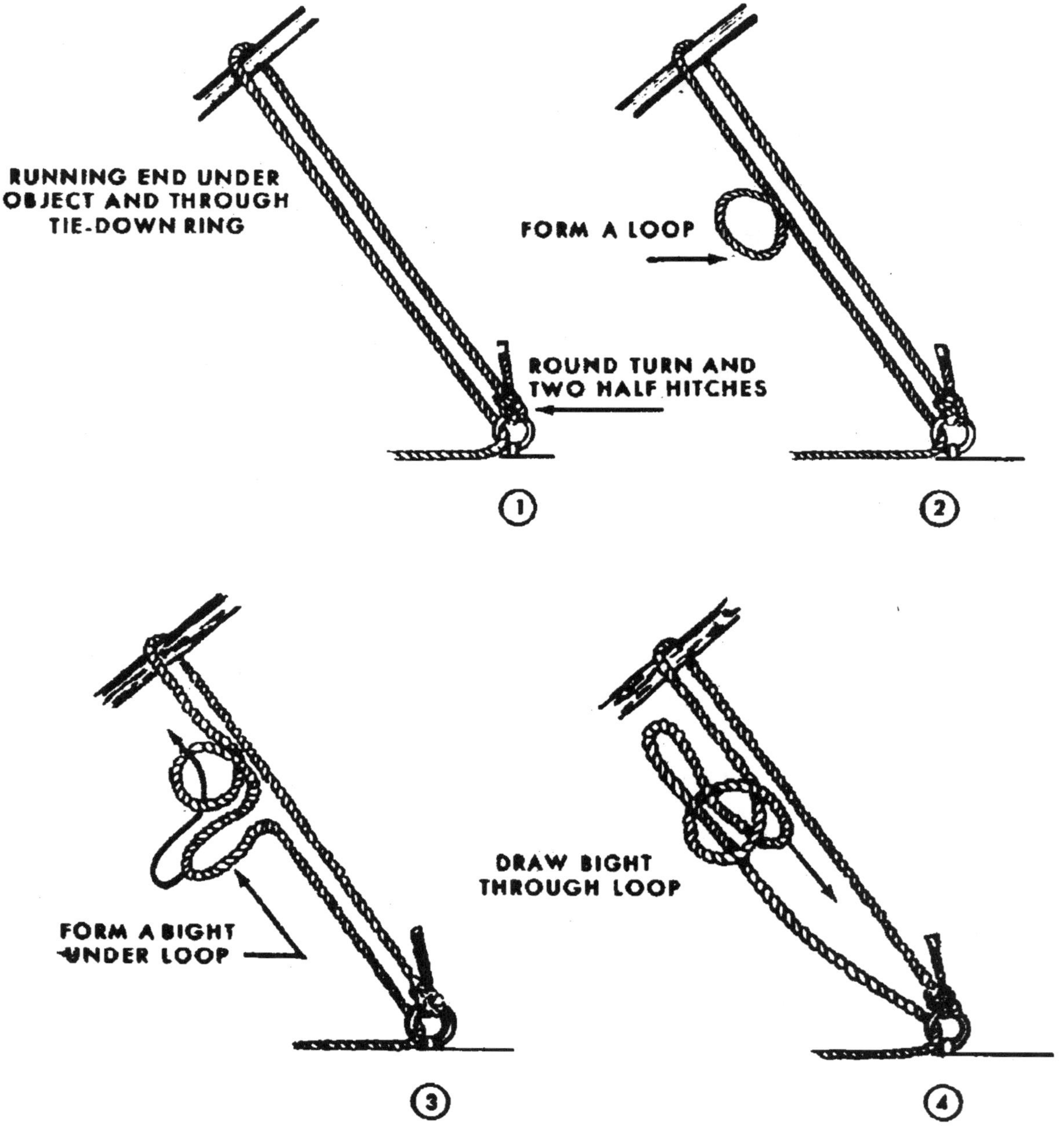

Figure 2-35. Baker bowline.

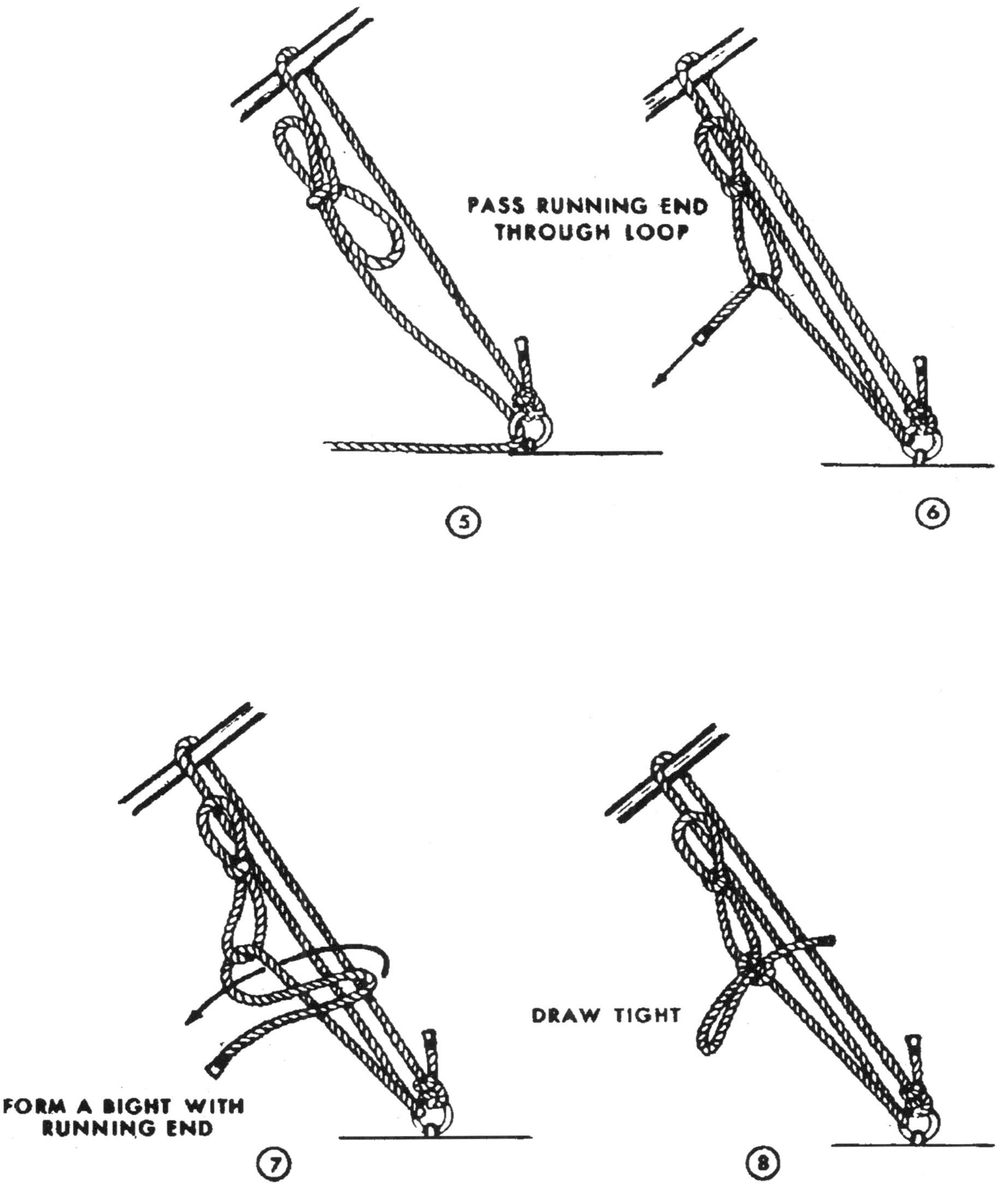

Figure 2-35—Continued.

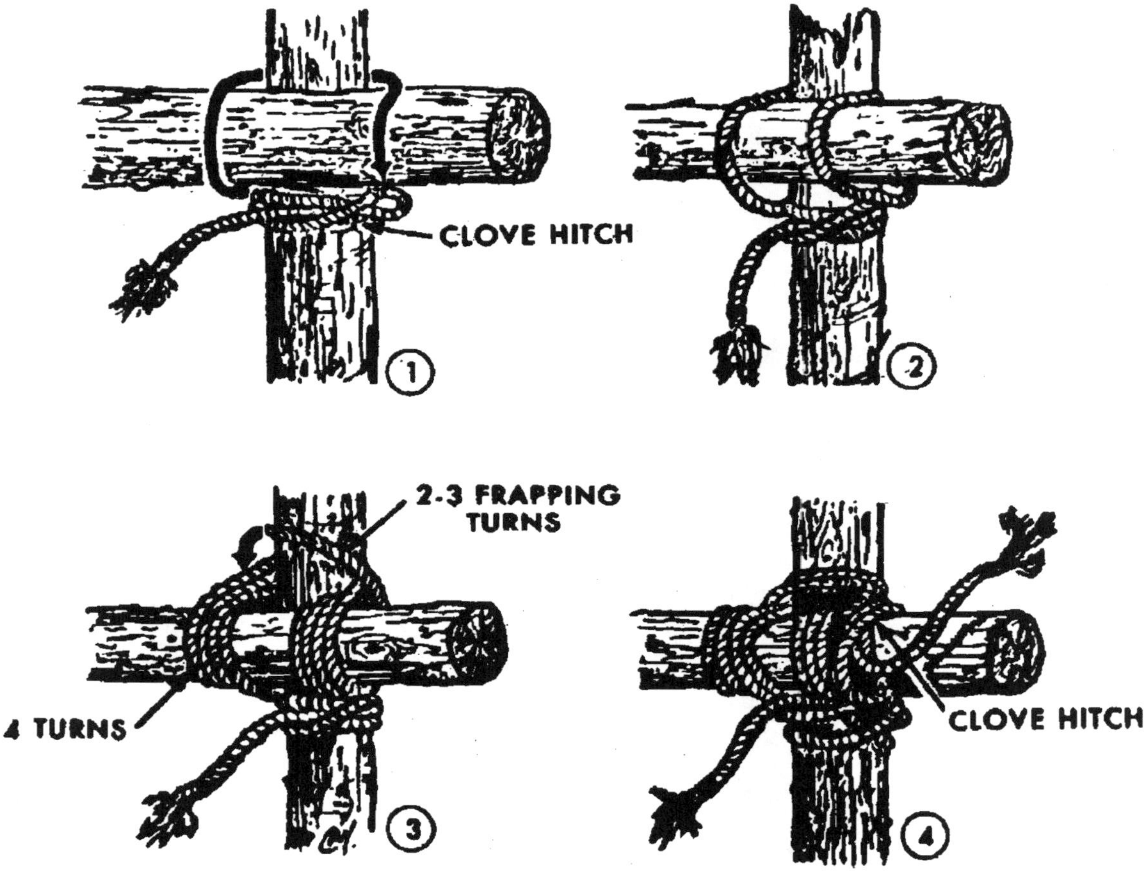

Figure 2–36. Square lashing.

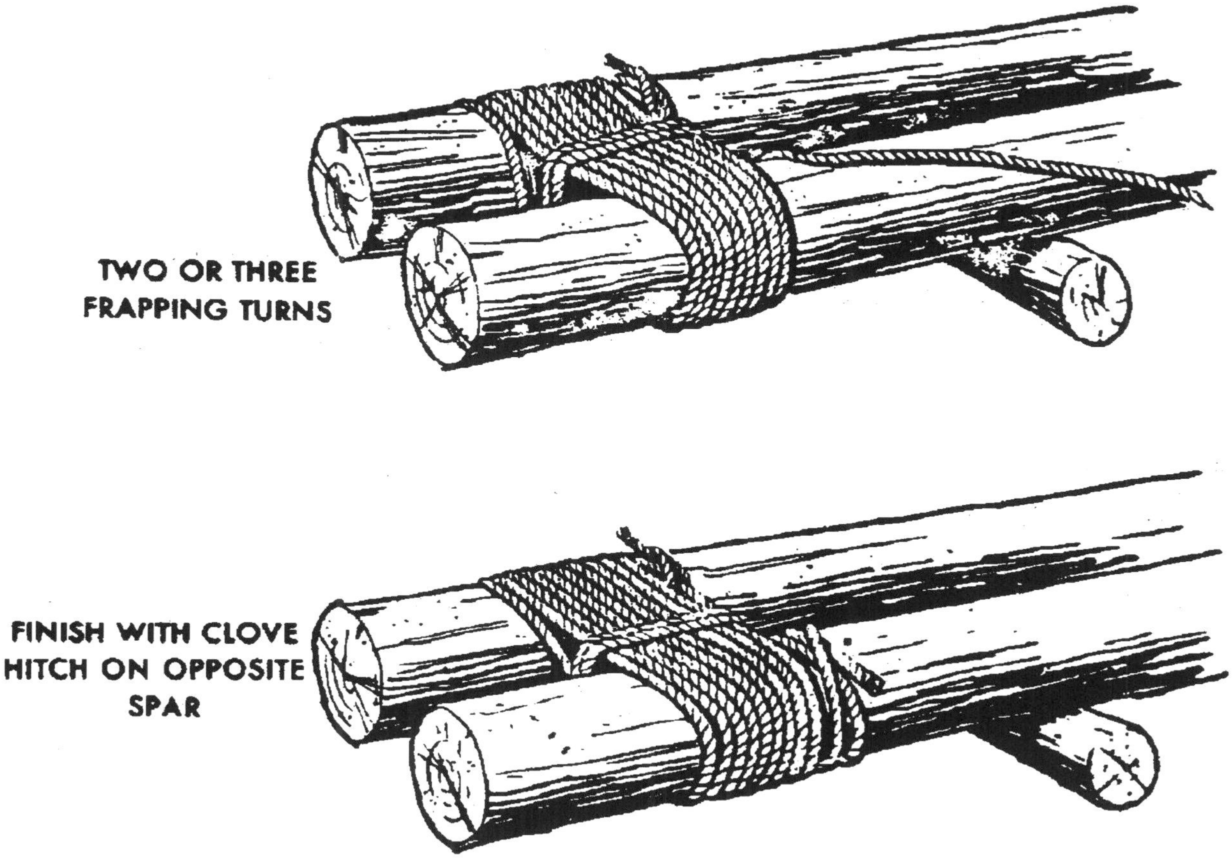

Figure 2–37. Shears lashing.

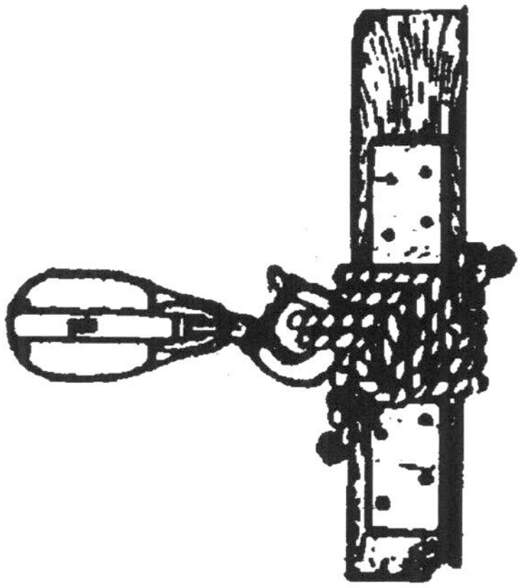

Figure 2–38. Block lashing.

Section II. SPLICES

2–9. Introduction

Splicing is a method of joining rope or wire by unlaying strands of both ends and interweaving these strands together. There are four general types of splices—a short splice, an eye or side splice, a long splice, and a crown or back splice. The methods of making all four types of splices are similar. They generally consist of three basic steps—unlaying the strands of the rope, placing the rope ends together, and interweaving the strands and tucking them into the rope. It is extremely important, in the splicing of wire rope, to use great care in laying the various rope strands firmly into position. Slack strands will not receive their full share of the load and cause excessive stress to be put on the other strands. The unequal stress distribution will decrease the possible ultimate strength of the splice. When splices are to be used in places where their failure may result in material damage or may endanger human lives, the splices should be tested under stresses equal to at least twice their maximum working load before the ropes are placed into service. Table 2–1 shows the amount or length of rope to be unlaid on each of the two ends of the ropes, and the amount of tuck for ropes of different diameters. As a rule of thumb use the following: long splice—40 times the diameter; short splice—20 times the diameter.

Table 2–1. Amount of Wire Rope To Allow for Splice and Tucks

Length of rope to allow in feet				Length of tuck in inches		
Diameter (inches)	Short splice	Eye splice	Long splice	Short splice	Eye splice	Long splice
¼–⅜	15	1	30	15	1	30
½–⅝	20	2	40	20	2	40
¾–⅞	24	2½	50	24	2½	50
1–1⅛	28	3	60	28	3	60
1¼–1⅜	32	3½	70	32	3½	70
1½	36	4	80	36	4	80

2–10. Short Splice for Fiber Rope

The short splice (fig. 2–39) is as strong as the rope in which it is made and will hold as much as a long splice. However, the short splice causes an increase in the diameter of the rope for a short distance and can be used only where this increase in diameter will not affect operations. It is called the short splice because a minimum reduction in rope length takes place in making the splice. This splice is frequently used to repair damaged ropes when two ropes of the same size are to be joined together permanently. Damaged parts of a rope are cut out and the sound sections are spliced.

2–11. Eye or Side Splice for Fiber Rope

The eye or side splice (fig. 2–40) is used for making a permanent loop in the end of a rope. The loops can be used for fastening the rope to a ring or hook and can be made up with or without a thimble. A thimble is used to reduce wear. This splice is also used to splice one rope into the side of another. As a permanent loop or eye, no knot can compare with this splice for neatness and efficiency.

2–12. Long Splice for Fiber Rope

The long splice (fig. 2–41) is used when the larger diameter of the short splice has an adverse effect on the use of the rope, and for splicing long ropes that operate under heavy stress. This splice is as strong as the rope itself. A skillfully made long splice will run through sheaves without any difficulty. The ropes to be joined should be the same lay and as nearly the same diameter as possible.

2–13. Crown or Back Splice for Fiber Rope

Where the end of a rope is to be spliced to prevent unlaying and a slight enlargement of the end is not objectionable, a crown splice (fig. 2–42) may be used to accomplish this. No length of rope should be put into service without having the ends properly prepared.

2–14. Renewing Strands

When one strand of a rope is broken it cannot be repaired by tying the ends together because this would shorten the strand. The rope can be repaired by inserting a strand longer than the break and tying the ends together (fig. 2–43).

2–15. Tools for Splicing

Only a few tools are required for splicing wire rope. In addition to the tools shown in figure 2–44, a hammer and cold chisel are often used for cutting ends of strands. Two slings of marline and two sticks should be used for untwisting the wire. A pocket knife may be needed for cutting the hemp core.

Figure 2–39. Short splice for fiber rope.

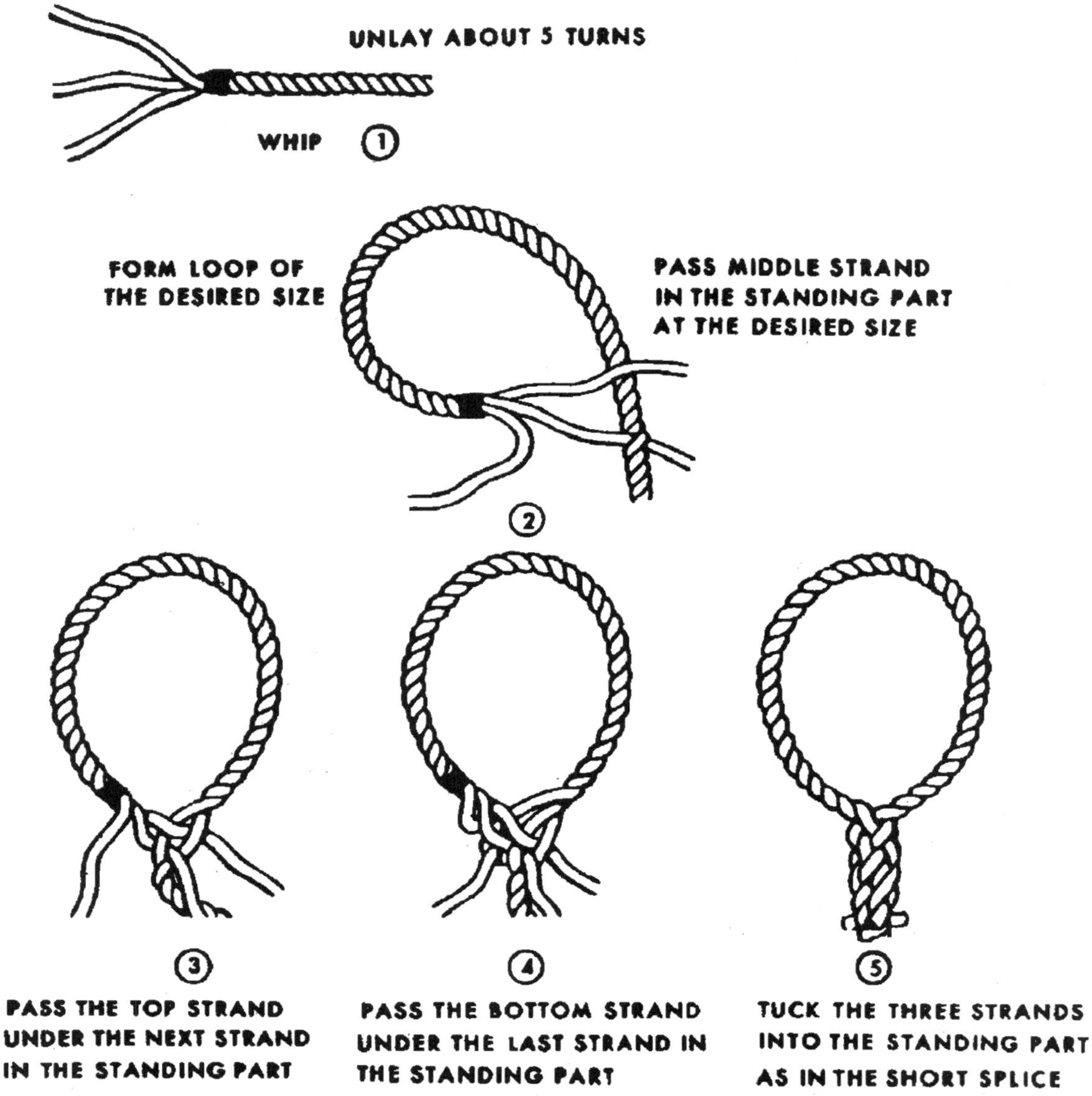

Figure 2–40. Eye or side splice for fiber rope.

2–16. Short Splice in Wire Rope

A short splice develops only from 70 to 90 percent of the strength of the rope. A short splice is bulky and used only for block straps, slings, or where an enlargement of the diameter is of no importance. It is not suitable for splicing driving ropes or ropes used in running tackles, and should never be put into a crane or hoist rope. The wire rope splice differs from the fiber rope splice (fig. 2–39) only in the method of tucking the end strands (fig. 2–45).

2–17. Eye Splice in Wire Rope

An eye splice can be made with or without a thimble. A thimble (fig. 2–46) should be used for every rope eye unless special circumstances prohibit it. The thimble protects the rope from sharp bends and abrasive action. The efficiency

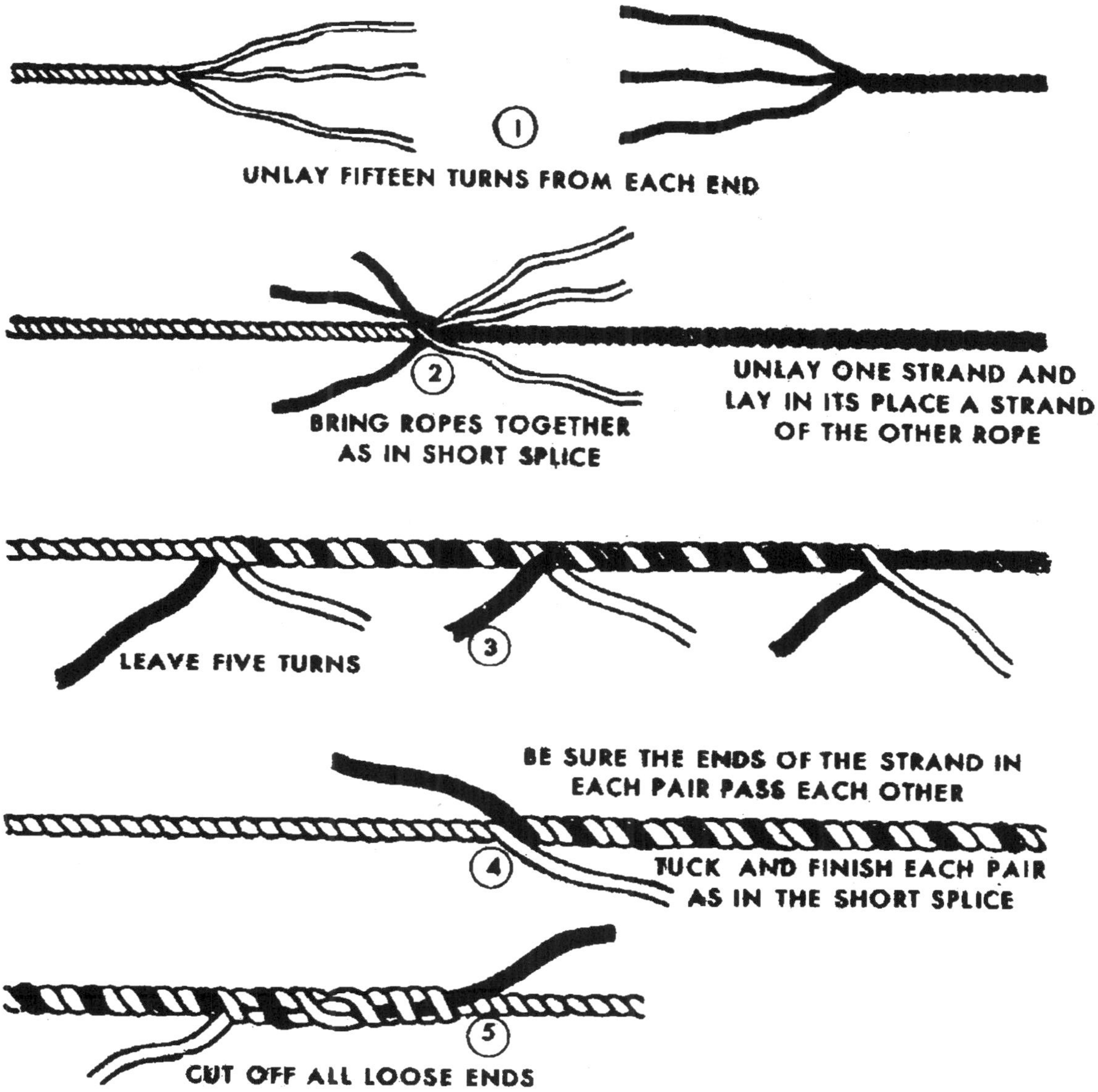

Figure 2–41. Long splice for fiber rope.

of a well-made eye splice with a heavy-duty thimble varies from 70 to 90 percent. Occasionally it becomes necessary to construct a field expedient, called hasty eye (fig. 2–47). The hasty eye can be easily and quickly made, but is limited to about 70 percent of the strength of the rope, and consequently should not be used for hoisting loads.

2–18. Long Splice in Wire Rope

The long splice (fig. 2–48) is used for joining two ropes or for making an endless sling without increasing the thickness of the wire rope at the splice. It is the best and most important kind of splice because it is strong and trim.

a. Round Strand Regular Lay Rope. The directions given in figure 2–48 are for making a

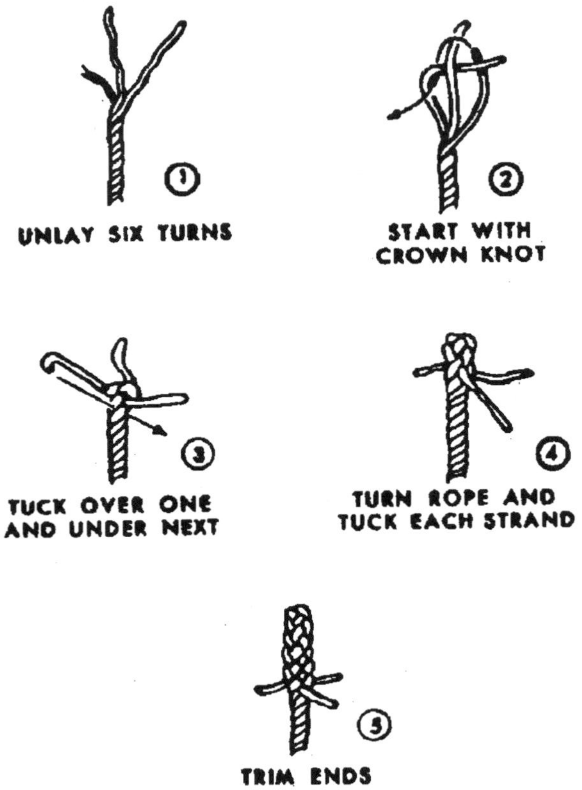

Figure 2 42. Crown or back splice for fiber rope.

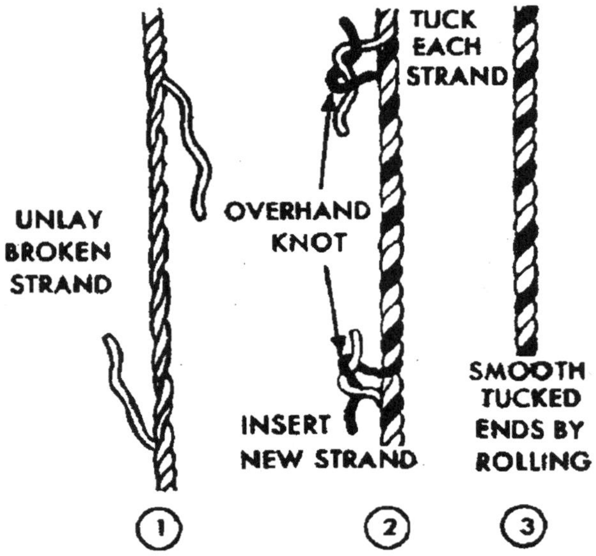

Figure 2 43. Renewing rope strands.

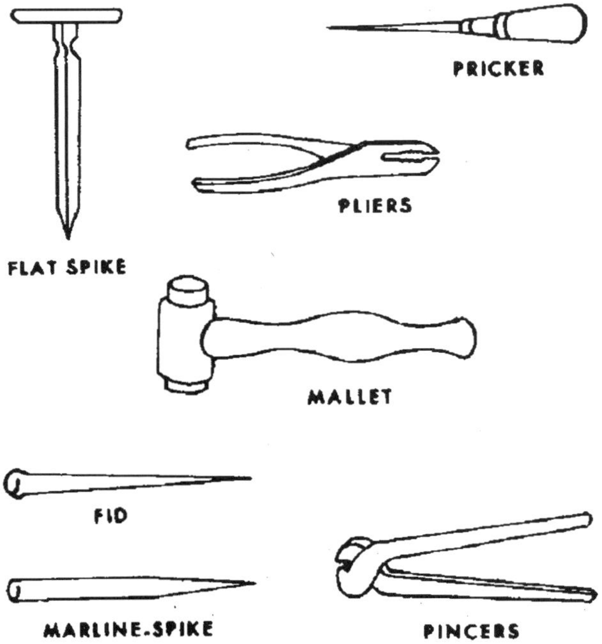

Figure 2–44. Tools for wire splicing.

30-foot splice in a ¾-inch regular lay, round strand, hemp center wire rope. Other strand combinations differ only when there is an uneven number of strands. In splicing ropes having an odd number of strands, the odd tuck is made at the center of the splice.

b. Round Strand Lang Lay Rope. In splicing a round strand Lang lay rope, it is advisable to make a slightly longer splice than for the same size rope of regular lay because of the tendency of the rope to untwist. Up to the point of tucking the ends, the procedure for regular lay is followed. Then, instead of laying the strands side-by-side where they pass each other, they are crossed over to increase the holding power of the splice. At the point where they cross, the strands are untwisted for a length of about 3 inches so they cross over each other without materially increasing the diameter of the rope. Then the tucks are finished in the usual manner.

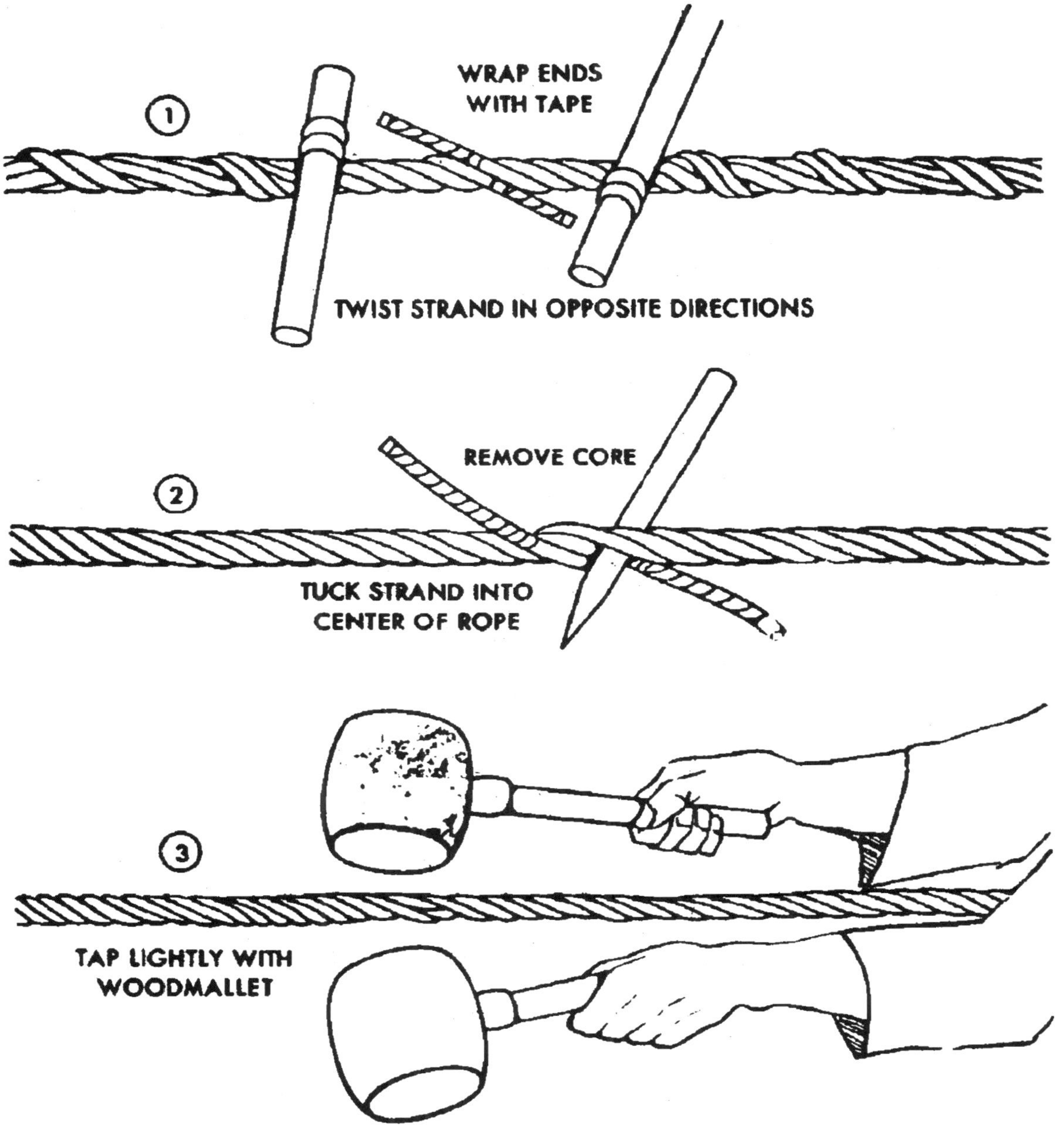

Figure 2-45. Tucking wire rope strands.

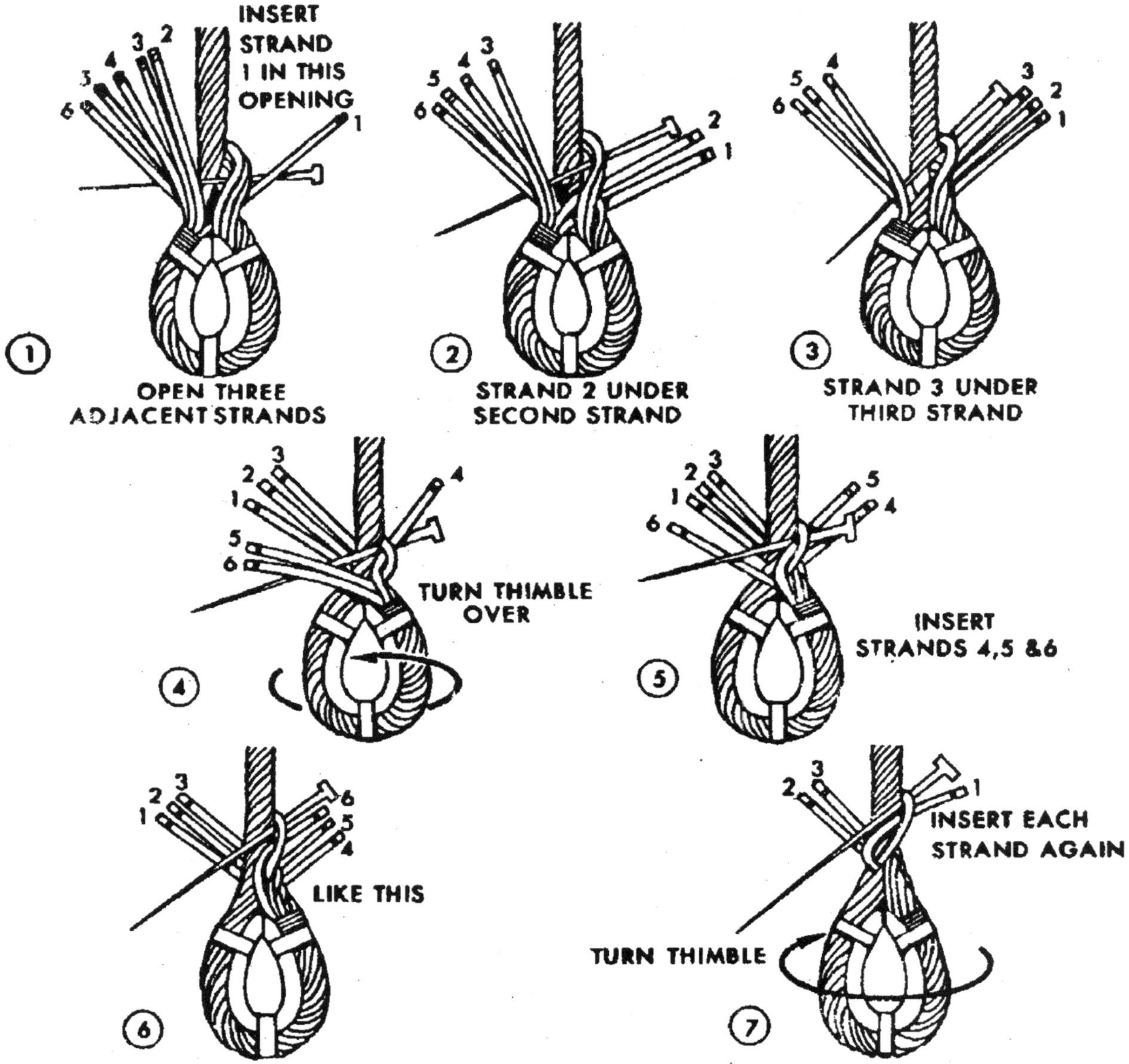

Figure 2–46. Eye splice with thimble.

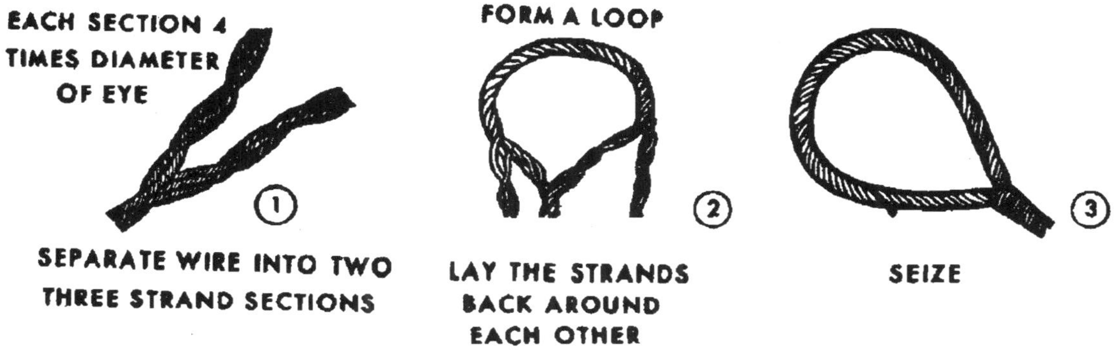

Figure 2–47. Hasty eye splice. (Use only preformed rope).

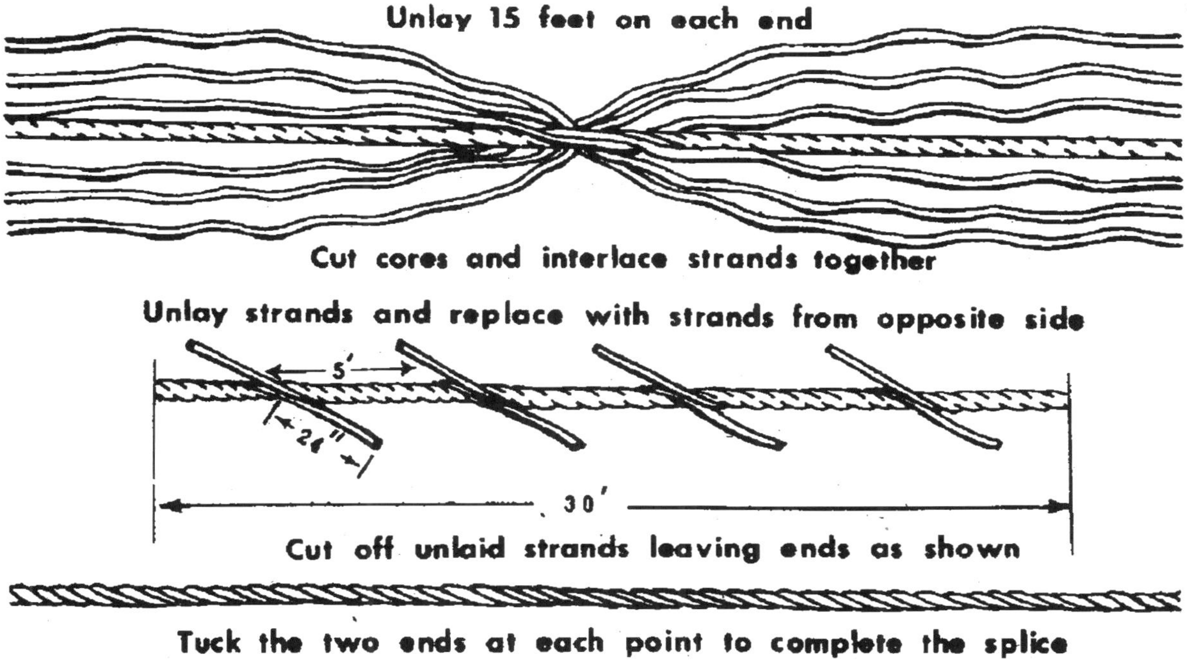

Figure 2–48. Long splice in wire rope.

Section III. ATTACHMENTS

2–19. Use of Attachments

Most of the attachments used with wire rope are designed to provide an eye on the end of the rope by which maximum strength can be obtained when the rope is connected with another rope, hook, or ring. Figure 2–49 shows a number of attachments used with the eye splice. Any two of the ends can be joined together, either directly or with the aid of a shackle or end fitting. These attachments for wire rope take the place of knots.

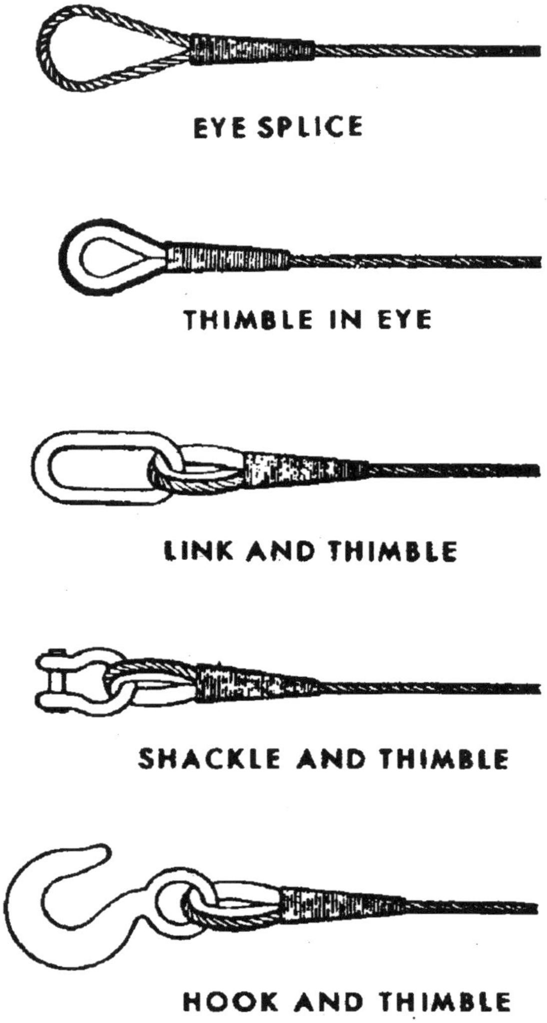

Figure 2 49. Attachments used with eye splice.

2–20. End Fittings

An end fitting may be placed directly on wire rope. Fittings that are easily and quickly changed are clips, clamps, and wedge sockets. The basket socket end fittings (fig. 2–50) include closed sockets, open sockets, and bridge sockets.

2–21. Clips

Wire rope clips (fig. 2–51) are reliable and durable. They can be used repeatedly in making eyes in wire rope, either for a simple eye or an eye reinforced with a thimble, or to secure a

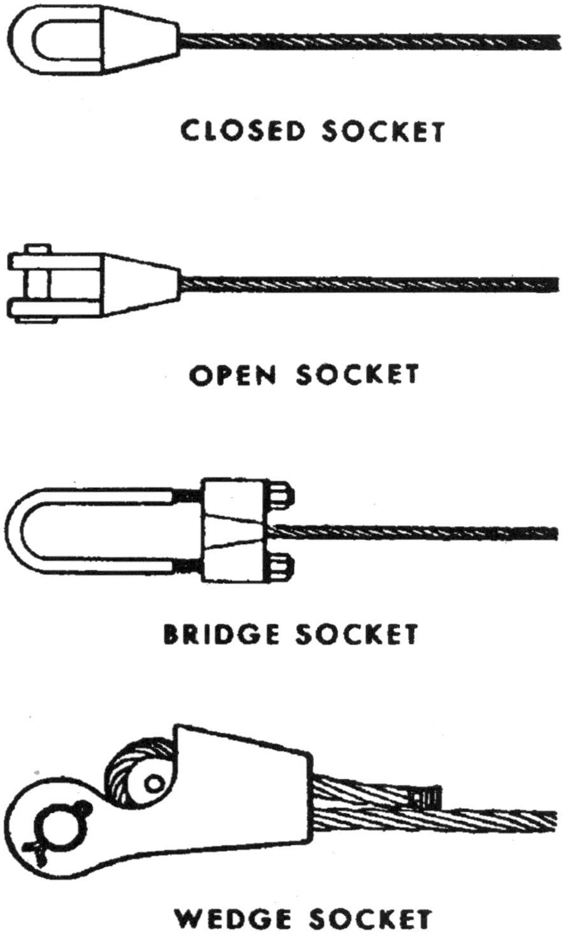

Figure 2–50. Basket socket end fittings.

wire rope line or anchorage. The clips should be spaced about six rope diameters apart. The number of clips to be installed is equal to three times the diameter of the rope plus one. (No. of clips = 3 D + 1) Thus, a 1-inch rope requires four clips. When this calculation results in a fraction the next larger whole number is used. After all clips are installed the clip farthest from the thimble is tightened with a wrench. Then the rope is placed under tension and the nuts are tightened on the clip next to the first clip. The remaining clips are tightened in order, moving toward the thimble. After the rope has been placed in service and has been under tension, the nuts should be tightened again to compensate for any decrease in rope diameter caused by the load. For this reason, clips should never be placed under ground.

2–22. Clamps

A wire clamp (fig. 2–52) can be used with or without a thimble to make an eye in wire rope. Ordinarily a clamp is used to make an eye without a thimble. It has about 90 percent of the strength of the rope. The two end collars should be tightened with wrenches to force the clamp to a good snug fit. This crushes the pieces of rope firmly against each other.

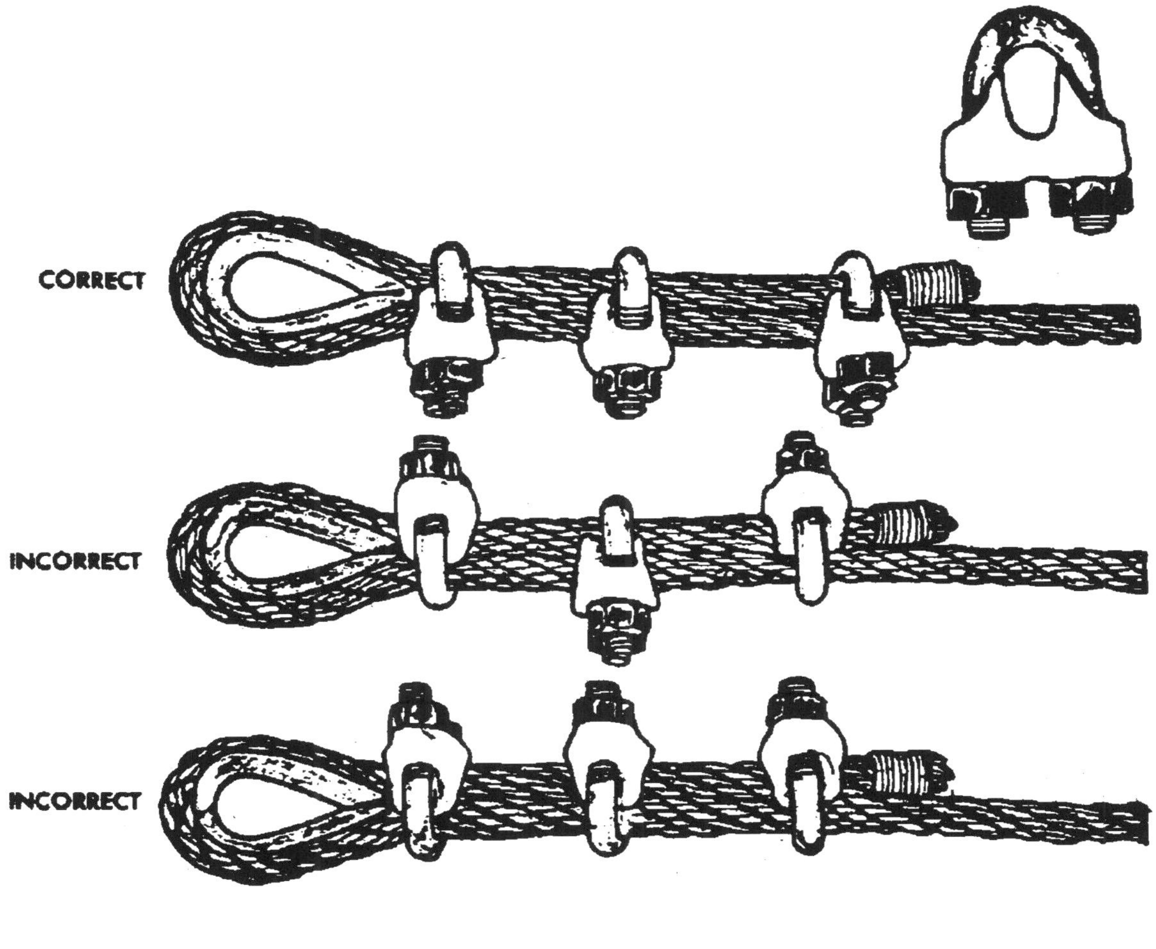

Figure 2–51. Wire rope clips.

2–23. Wedge Socket

A wedge socket end fitting (fig. 2–53) is used when it may be necessary to change the fitting at frequent intervals. The efficiency is about two-thirds the strength of the rope. It is made in two parts. The socket itself has a tapered opening for the wire rope and a small wedge to go into this tapered socket. The loop of wire

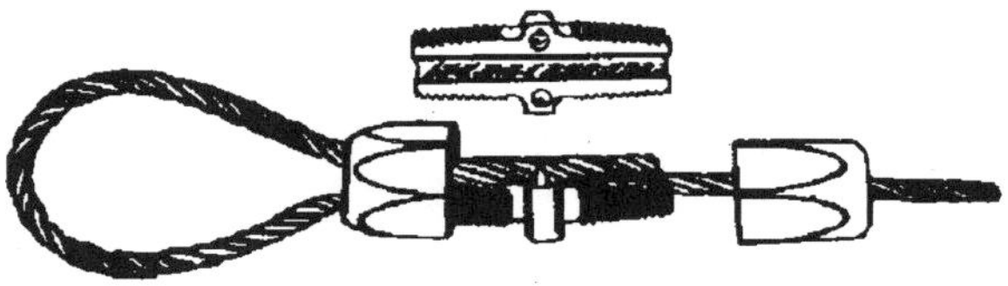

Figure 2–52. Wire rope clamps.

rope must be inserted in the wedge socket so that the standing part of the wire rope will form a nearly direct line to the clevis pin of the fitting. A properly installed wedge socket connection will tighten when a strain is placed on the wire rope.

2–24. Basket Socket

A basket type socket ordinarily is attached to the end of the rope with molten zinc or babbitt

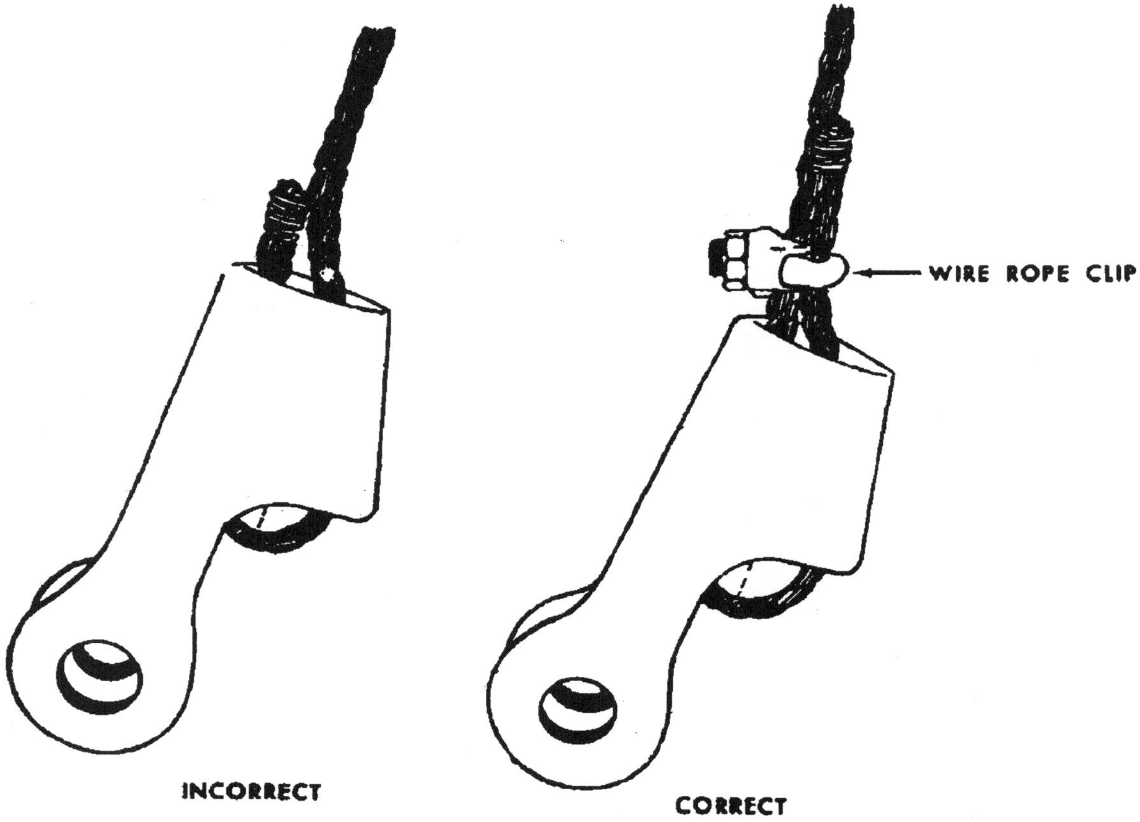

Figure 2–53. Wedge socket.

metal, and is a permanent end fitting. If properly made up, this fitting is as strong as the rope itself. If molten lead is used instead of zinc, the strength of the connection must be assumed to be reduced to one-fourth the strength of a zinc connection. The socket can be made up by the dry method if facilities are not available to make a poured fitting, but its strength is sharply reduced and must be considered to be about one-sixth the strength of a zinc connection. In all cases the wire rope should lead from the socket in line with the axis of the socket.

a. Poured Method. The poured basket socket (fig. 2–54) is the most satisfactory method in use. If the socketing is properly done, when tested to destruction, a wire rope will break before it will pull out from the socket.

b. Dry Method. The dry method (fig. 2–55) should be used only when facilities are not available for the poured method. The strength of the connection must be assumed to be reduced to about one-sixth of the strength of a poured zinc connection.

2–25. Stanchions

The standard pipe stanchion (fig. 2–56) is made up of a 1-inch diameter pipe. Each stanchion is 40 inches long. Two ¾-inch wire rope clips are fastened through holes in the pipe with the centers of the clips 36-inches apart. Such a stanchion can be used without modification for a suspended walkway which uses two wire ropes on each side, but for handlines the lower wire rope clip is removed or left off. Refer to TM 5–270 for detailed information on types and uses of stanchions.

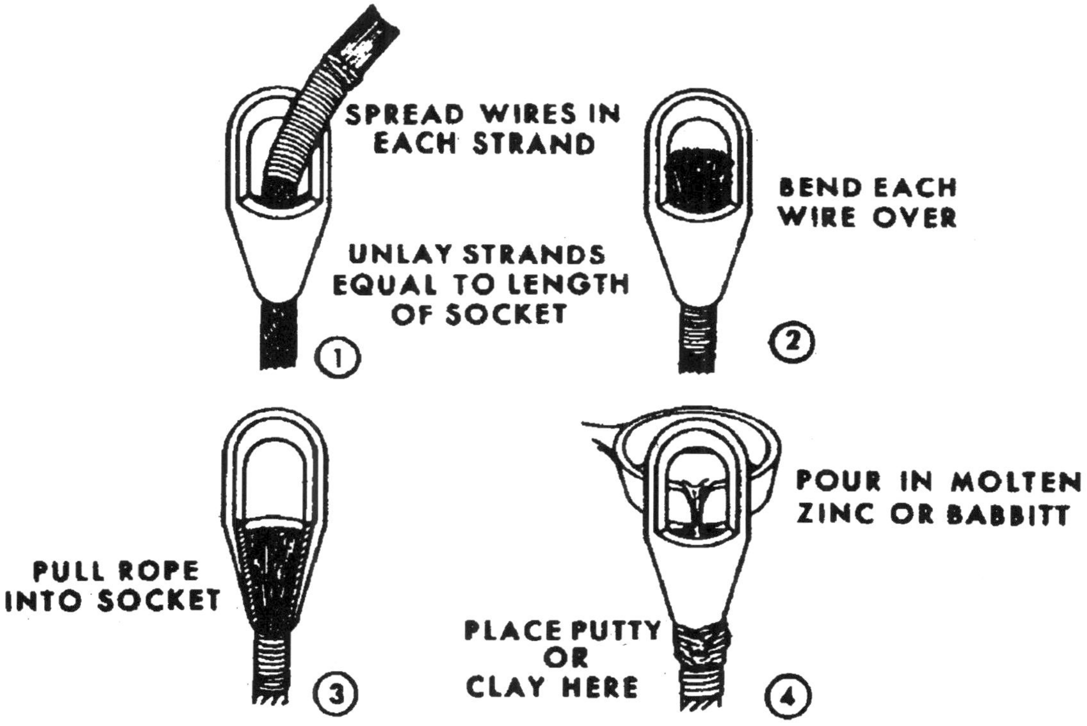

Figure 2–54. Attaching basket sockets by pouring.

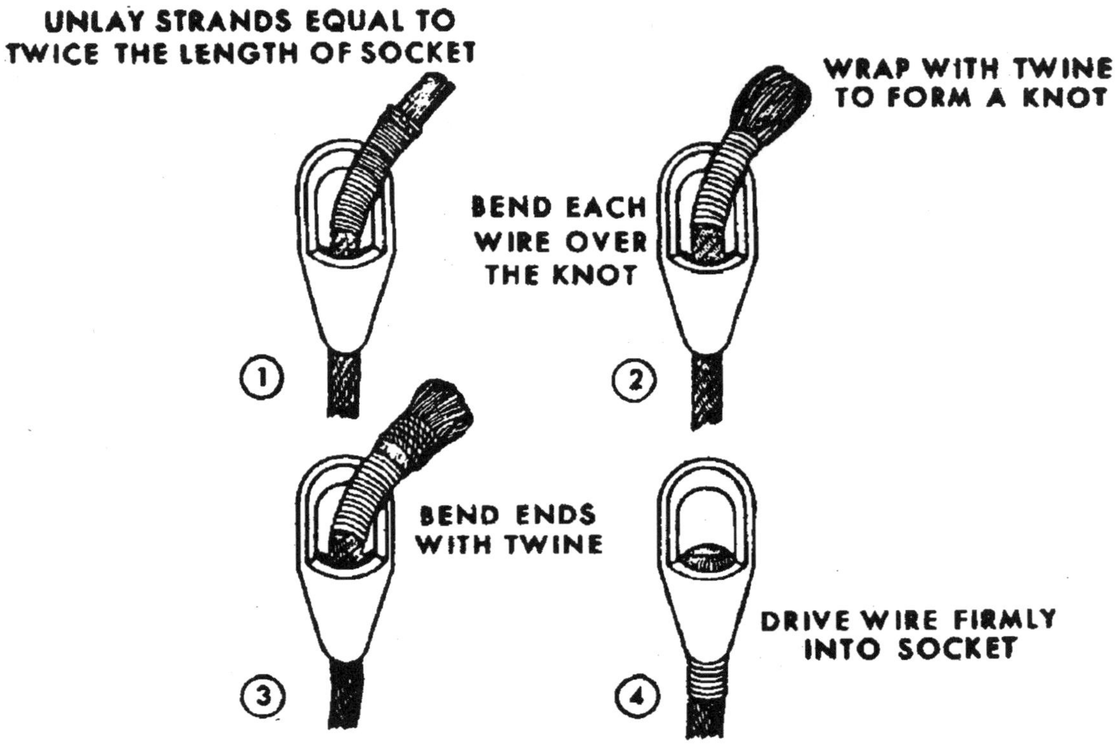

Figure 2–55. Attaching basket socket by dry method.

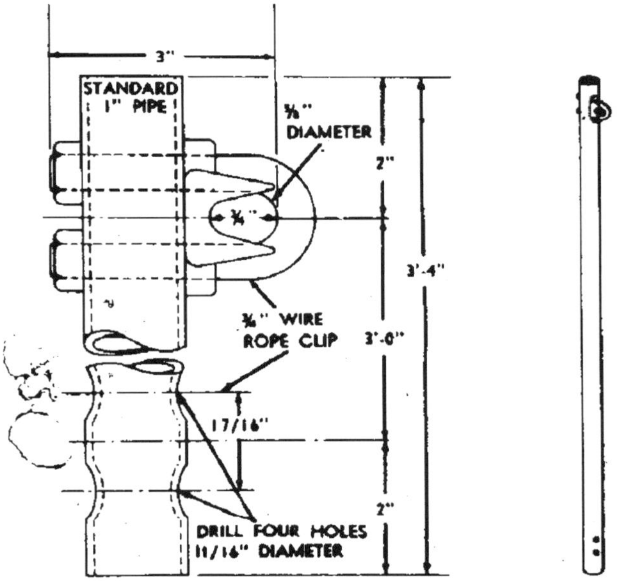

Figure 2–56. Iron pipe stanchions.

2–26. Use of Ropes

Ropes may be used in the construction of hanging ladders and standoff ladders. Hanging ladders are made of wire or fiber rope anchored at the top and suspended vertically. They are difficult to ascend and descend, particularly for a man carrying a pack or load, and should be used only when necessary. Standoff ladders are easier to climb because they have two wood or metal uprights which hold them rigid, and they are placed at an angle. Both types of ladders can be prefabricated and transported easily. One or two standoff ladders are adequate for most purposes, but three or four hanging ladders must be provided for the same purpose because they are more difficult to use.

2–27. Hanging Ladders

The uprights of hanging ladders may be made of wire or fiber rope and anchored at the top and bottom. Wire rope uprights with pipe rungs make the most satisfactory hanging ladders because they are more rigid and do not sag as much as hanging ladders made of other material. Wire rope uprights with wire rope rungs are usable. Fiber rope uprights with wood or fiber rope rungs are difficult to use because of their greater flexibility which causes them to twist when they are being used. A log should be placed at the break of the ladder at the top to hold the uprights and rungs away from a rock face so that better handholds and footholds are provided. A single rock anchor is usually sufficient at the bottom of the ladder, or a pile of rocks can be used as bottom anchor for fiber rope hanging ladders.

a. Wire Rope Ladder With Pipe Rungs. A wire rope ladder can be made using either 1-inch or ¾-inch pipe rungs. The 1-inch pipe rungs are more satisfactory. For such ladders the standard pipe stanchion is used. The pipe stanchions are spaced 12 inches apart in the ladder (fig. 2–57) and the ¾-inch wire rope clips are inserted in the stanchion over ¾-inch wire rope uprights. If ⅜-inch wire rope uprights are used, ⅜-inch wire rope clips are inserted in the pipe over the wire rope uprights. When ¾-inch pipe rungs are used, the rungs are also spaced 12 inches apart in the ladder but uprights should not be spaced more than 12 inches apart because of the weaker pipe used. The rungs may be fastened in place by two different methods. In one method a 7/16-inch diameter hole is drilled at each end of each pipe rung and ⅜-inch wire rope uprights are threaded through the holes. To hold each rung in place a ⅜-inch wire rope clip is fastened about the wire rope upright at each end of each rung after the rung is in final position. In the other method the pipe rungs are cut 12 inches long and the U-bolt of a ⅜-inch rope clip is welded to each end. The rungs are spaced 12 inches apart on the ⅜-inch wire rope uprights. The saddle of the wire rope clips and the nuts are placed on the U-bolts, then the nuts are tightened to hold the rungs in place.

b. Wire Rope Ladder With Wire Rope Rungs. A wire rope ladder with wire rope rungs is made by laying the ⅜-inch diameter wire rope unrights on the ground. The first length is layed out in a series of U-shaped bends. The second length is layed out in a similar manner (fig. 2–58) with the U-shaped bends in the opposite direction from those in the first series, and the horizontal rung portions overlapping. A ⅜-inch wire rope clip is fastened on the overlapping rung portions at each end of each rung to hold them firm.

c. Fiber Rope Ladder With Fiber Rope Rungs. Fiber rope ladders with fiber rope rungs can be made by using two or three uprights. When three uprights (fig. 2–59) are used, a loop is made in the center upright at the position of each rung. The two outside uprights are spaced 20 inches apart. A loop and a single splice hold each end of each rung to the outside upright. A loop in the center of the rung passes through the loop in the center upright. If only two uprights are used, the rungs are held in place by a loop and a rolling hitch or a single splice at each upright. The two uprights must be closer together, with shorter rungs, to stiffen the ladder. Ladders of either type are very flexible and difficult to climb.

d. Fiber Rope Ladder With Wood Rungs. Fiber rope ladders with wood rungs (fig. 2–60) can be made by using finished lumber or native material for rungs. When native material is

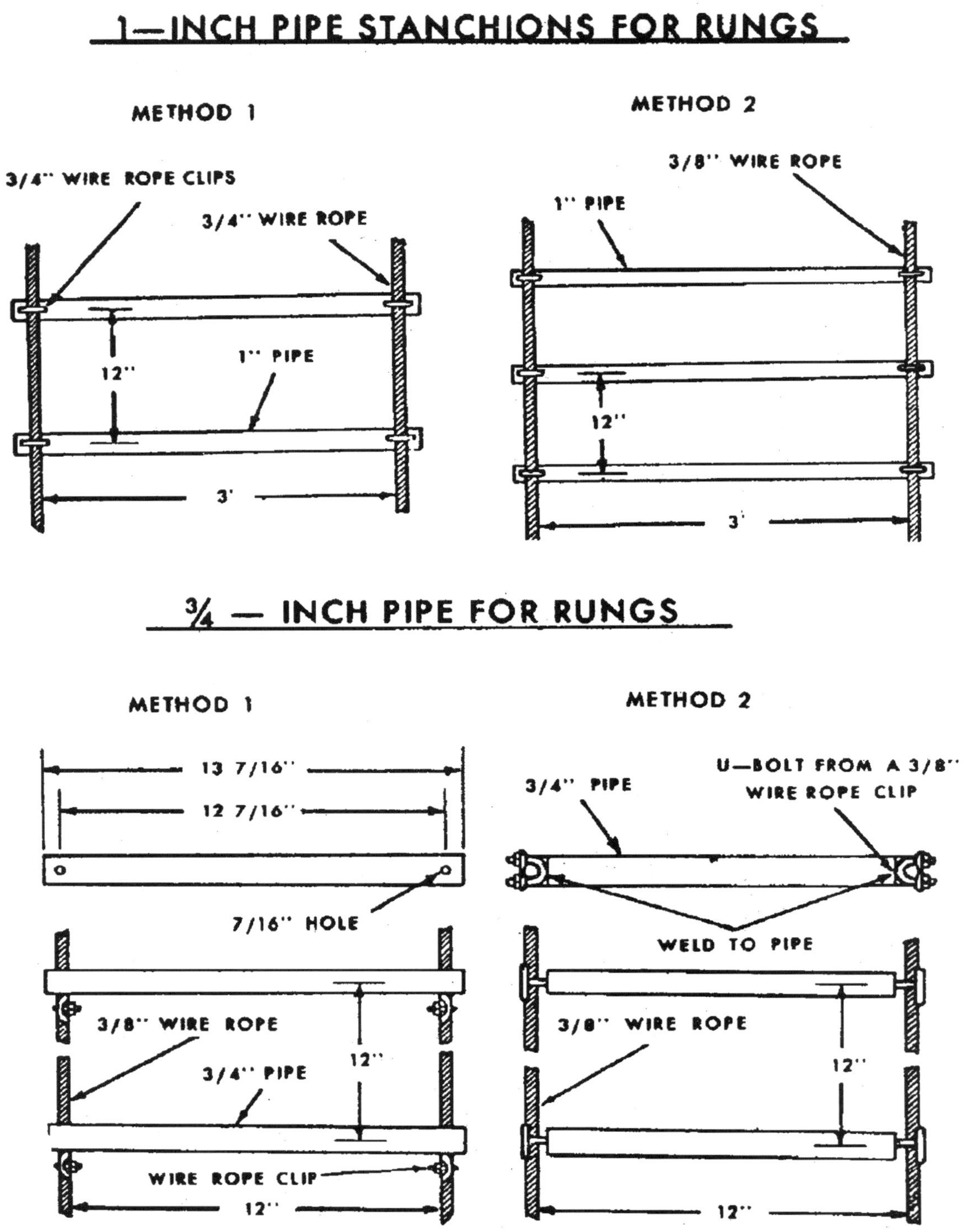

Figure 2–57. Pipe rungs.

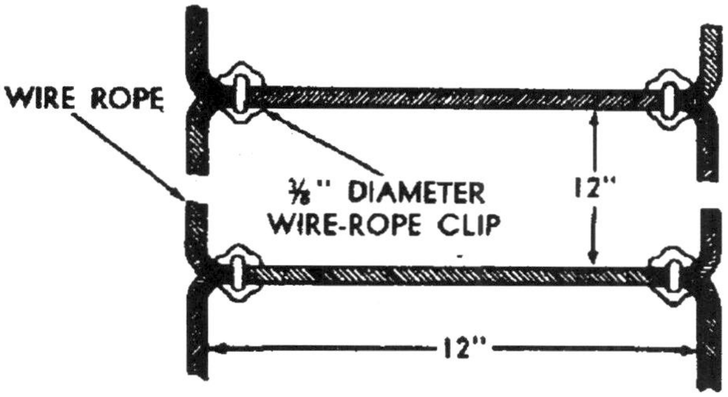

Figure 2–58. Wire rope rungs.

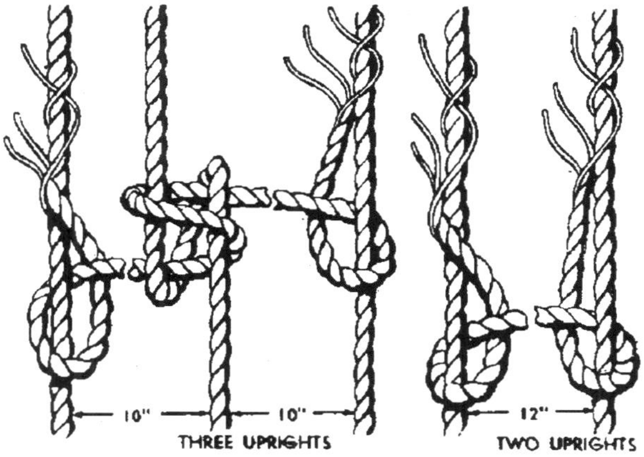

Figure 2–59. Fiber rope rungs.

used, the rungs are cut from 2-inch diameter material about 15 inches long. The ends of each rung are notched and the rung is fastened to the fiber rope upright with a clove hitch. The rungs are spaced 12 inches apart. A piece of seizing wire is twisted about the back of the clove hitch to make it more secure, and in a manner which will not snag the clothing of persons climbing the ladder. If rungs are to be made of finished lumber the rungs are cut to size and a ¾-inch hole is drilled at each end. Oak lumber is best for this purpose. A ¼-inch by 2½-inch carriage bolt is put horizontally through each end near the vertical hole to prevent splitting. An overhand knot is tied in the upright to support the rung. Then the upright is threaded through the ¾-inch hole in the rung. A second overhand knot is tied in the upright before it is threaded through the next rung. This procedure is continued until the desired length of the ladder is reached.

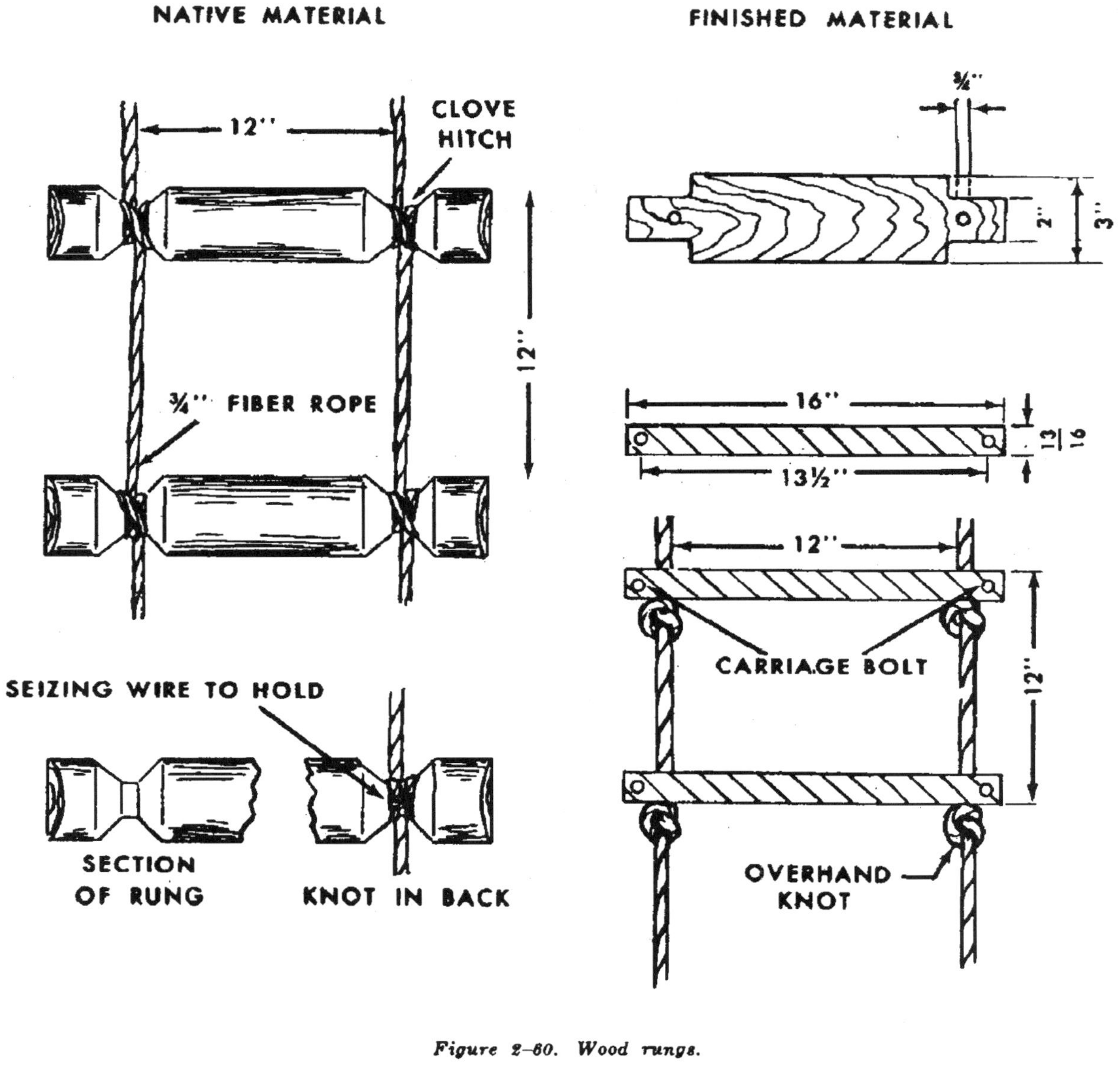

Figure 2–60. Wood rungs.

CHAPTER 3

HOISTING

Section I. CHAINS AND HOOKS

3–1. Introduction

Chains are made up of a series of links fastened through each other. Each link is made of a rod of wire bent into an oval shape and welded at one or two points. The weld ordinarily causes a slight bulge on the side or end of the link (fig. 3–1). The chain size refers to the diameter in inches of the rod used to make the link. Chains will usually stretch under excessive loading so that the individual links will be bent slightly. Bent links are a warning that the chain has been overloaded and might fail suddenly under load. Wire, on the other hand, will fail a strand at a time, giving warning before complete failure occurs. If a chain is equipped with the proper hook, the hook should start to fail first, indicating that the chain is overloaded. Chains are much more resistant to abrasion and corrosion than wire rope; therefore, chains are used where this type of deterioration is a problem. An example is the use of chains for anchor gear in marine work where the chains must withstand the corrosive effects of sea water. Another example is the use of chains for slings to lift heavy objects with sharp edges which would cut wire.

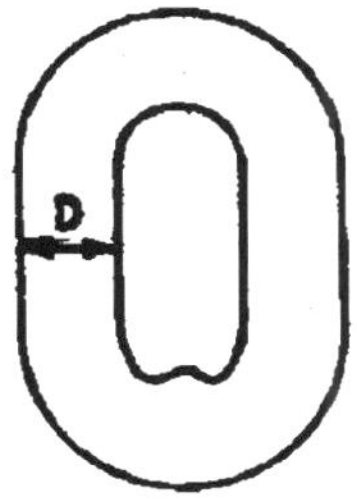

Figure 3–1. Link thickness.

A number of grades and types of chains are available. In lifting, chains as well as fiber ropes or wire ropes can be tied to the load. But for speed and convenience, it is much better to fasten a hook to the end of the lifting line. Blocks are ordinarily constructed with a hook (para 3–4).

3–2. Strength of Chains

To determine the safe working load on a chain, apply a factor of safety to the breaking strength. The safe working load ordinarily is assumed to be approximately one-sixth of the breaking strength, giving a factor of safety of 6. Table 3–1 lists safe working loads for various chains. The safe load or safe working capacity of an open link chain can be approximated by using the following rule of thumb: $SWC = 8D^2$

D = Smallest link thickness or least diameter measured in inches (fig. 3–1)

SWC = Safe working capacity in tons.

EXAMPLE: Using the rule of thumb, the safe working capacity of a chain with a link thickness of ¾-inch is

$SWC = 8D^2 = 8(3/4)^2 = 4.5$ tons or 9,000 pounds.

The figures given assume that the load is applied in a straight pull rather than by an impact. An impact load occurs when an object is dropped suddenly for a distance and stopped. The impace load in such a case is several times the weight of the load.

3–3. Care of Chains

When hoisting heavy metal objects using chains for slings, padding should be inserted around the sharp corners of the load to protect

Table 3–1. *Properties of Chains (Factor of Safety 6)*

Size*	Approximate weight per linear foot in pounds	Safe working load in pounds			
		Common iron	High grade iron	Soft steel	Special steel
1/4	0.8	512	563	619	1,240
3/8	1.7	1,350	1,490	1,650	3,200
1/2	2.5	2,250	2,480	2,680	5,250
5/8	4.3	3,470	3,810	4,230	7,600
3/4	5.8	5,070	5,580	6,000	10,500
7/8	8.0	7,000	7,700	8,250	14,330
1	10.7	9,300	10,230	10,600	18,200
1 1/8	12.5	9,871	10,858	11,944	21,500
1 1/4	16.0	12,186	13,304	14,634	26,300
1 3/8	18.3	14,717	16,188	17,807	32,051

*Size listed is the diameter in inches of one side of a link.

the chain links from being cut. The padding may be either planks or heavy fabric. Chains should not be permitted to twist or kink when under strain, Links of chain should never be fastened together with bolts or wire because such connections weaken the chain and limit its safe working capacity. Worn or damaged links should be cut out of the chain and replaced with a cold shut link. The cold shut link must be closed and welded to equal the strength of the other links. The smaller chain links can be cut with a bolt cutter. Large chain links must be cut with a hacksaw or oxyacetylene torch. Chains must be inspected frequently, depending on the amount of use. Painting a chain to prevent rusting is not advisable because the paint will interfere with the freedom of action of the links. A light coat of lubricant can be applied to prevent rusting. Chains should be stored in a dry and well ventilated place to prevent rusting.

3–4. Hooks

There are two general types of hooks available, the slip hook and the grab hook (fig. 3–2). Slip hooks are made so that the inside curve of the hook is an arc of a circle, and may be used with wire rope, chains, or fiber rope. Chain links can slip through a slip hook so the loop formed in the chain will tighten under a load. Grab hooks have an inside curve which is nearly U-shaped so the hook will slip over a link of chain edgeways but will not permit the next link to slip through. Grab hooks have a more limited range of use than slip hooks. They are used on chains when the loop formed with the hook is not intended to close up around the load.

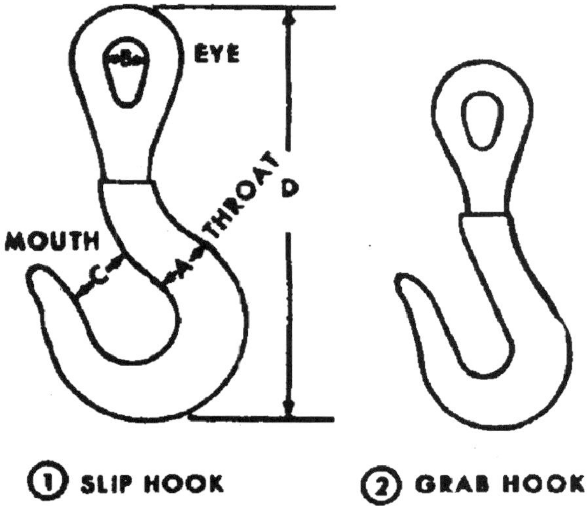

Figure 3–2. Types of hooks.

a. Strength. Hooks usually fail by straightening. Any deviation from the original inner arc indicates that the hook has been overloaded. Since evidence of overloading the hook is easily detected, it is customary to use a hook weaker than the chain to which it is attached. With this system, distortion of the hook will occur before the chain is overloaded. Severely distorted, cracked, or badly worn hooks are dangerous and should be discarded. Table 3–2 lists safe working loads on hooks. The safe working capacity of a hook can be approximated by using the following rule of thumb: $SWC = D^2$. D is the diameter in inches of the hook where the inside of the hook starts its arc (fig. 3–3). Thus, the safe working capacity of a hook with a diameter of 1¼ inches is as follows:

$$SWC = D^2 = (1\tfrac{1}{4})^2 = \frac{25}{16} \text{ tons or } 3125 \text{ pounds.}$$

b. Mousing. In general, a hook should always be "moused" as a safety measure to prevent slings or ropes from jumping off. Mousing also helps prevent straightening of the hook, but does not strengthen it materially. To mouse a hook (fig. 3–4) after the sling is on the hook, the wire or heavy twine is wrapped 8 or 10 turns around the two sides of the hook. The process is completed by winding several turns of the wire or twine around the sides of the mousing and tying the ends securely.

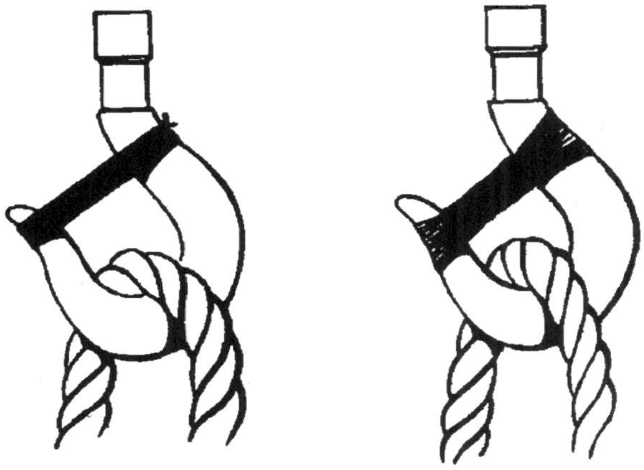

Figure 3–3. Hook thickness (diameter).

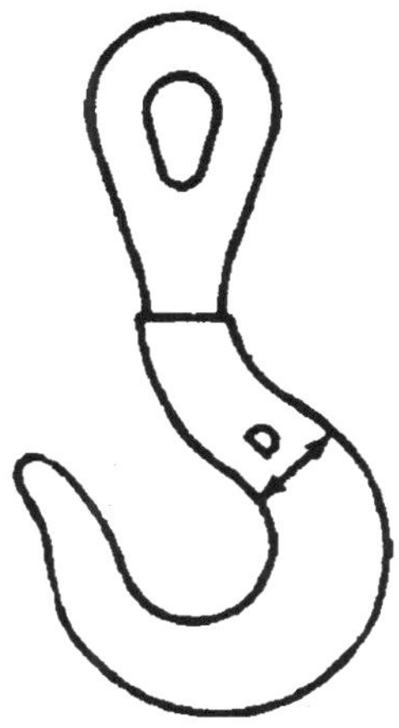

Figure 3–4. Mousing hooks.

Table 3 2. Safe Loads on Hooks

Diameter of metal A,* in.	Inside diameter of eye B, in.	Width of opening C, in.	Length of hook D, in.	Safe Working Capacity of hooks, lb.
11/16	7/8	1 1/16	4 15/16	1,200
3/4	1	1 1/8	5 13/32	1,400
7/8	1 1/8	1 1/4	6 1/4	2,400
1	1 1/4	1 3/8	6 7/8	3,400
1 1/8	1 3/8	1 1/2	7 5/8	4,200
1 1/4	1 1/2	1 11/16	8 19/32	5,000
1 3/8	1 5/8	1 7/8	9 1/2	6,000
1 1/2	1 3/4	2 1/16	10 11/32	8,000
1 5/8	2	2 1/4	11 21/32	9,400
1 7/8	2 3/8	2 1/2	13 9/32	11,000
2 1/4	2 3/4	3	14 13/16	13,600
2 5/8	3 1/8	3 3/8	16 1/2	17,000
3	3 1/2	4	19 3/4	24,000

*For reference to A, B, C, or D, see figure 3–2.

3–5. Inspection of Chains and Hooks

Chains, including the hooks, should be inspected at least once a month, but those that are used for heavy and continuous loading require more frequent inspections. Particular attention must be given to the small radius fillets at the neck of hooks for any deviation from the original inner arc. Each link and hook must also be examined for small dents, cracks, sharp nicks or cuts, worn surfaces, and distortions. Those that show any of these weaknesses must be replaced. If several links are stretched or distorted, the chain should not be used because it probably was overloaded or hooked improperly which weakened the entire chain.

Section II. SLINGS

3–6. Characteristics

The term "sling" includes a wide variety of designs. Slings may be made up of fiber rope, wire rope, or chain. The sling for lifting a given load may be an endless sling, a single sling, or several single slings used together to form a combination sling. The ends of single slings usually are made up into eyes, either with or without thimbles, to go over the hoisting hook. They may also be made up with end fittings to provide variable service. Spreaders may be added to change the angle of the sling legs. Each type or combination has its particular advantages which must be considered when selecting a sling for a given purpose. Fiber ropes make good sling material because of their flexibility, but they are more easily damaged by sharp edges on the material hoisted than are wire ropes or chain slings. Wire ropes are widely used for slings because they have a combination of strength and flexibility. Chain slings are used especially where sharp edges of metal would cut wire rope or where very hot items are lifted as in foundries or blacksmith shops. Fiber rope slings are used for lifting comparatively light loads and for temporary jobs. Properly designed and appropriately fabricated wire rope slings are the safest type of slings. They do not wear away as do slings made of fiber rope, nor do they lose their strength from exposure as rapidly. They also are not susceptible to the "weakest link" condition of chains caused by the uncertainty of the strengths of the welds. The appearance of broken wires clearly indicates the fatigue of the metal, and the end of the usefulness of the sling.

3–7. Types

a. Endless Slings. The endless sling is made by splicing the ends of a piece of wire rope or fiber rope together, or by inserting a cold shut link in a chain. Cold shut links should be welded after insertion in the chain. These endless slings are simple to handle, and may be used in several different ways to lift loads (fig. 3–5). A common method of using an endless sling is to cast the sling under the load to be lifted and inserting one loop through the other and over the hoisting hook. Such a sling is known as a choker hitch, or anchor hitch. When the hoisting hook is raised, one side of the choker hitch is forced down against the load by the strain on the other side, forming a tight grip on the load. If the endless sling is passed around the object to be lifted and both remaining loops are slipped over the hook, it is called a basket hitch. The inverted basket hitch is very much like the simple basket hitch, except that the two parts of the sling going under the load are spread wide apart. The toggle hitch is used only for special applications. It is actually a modification of the inverted basket hitch, except that the line passes around toggles fastened to the load rather than going around the load itself. The barrel slings can be made with fiber rope to hold barrels horizontally or vertically.

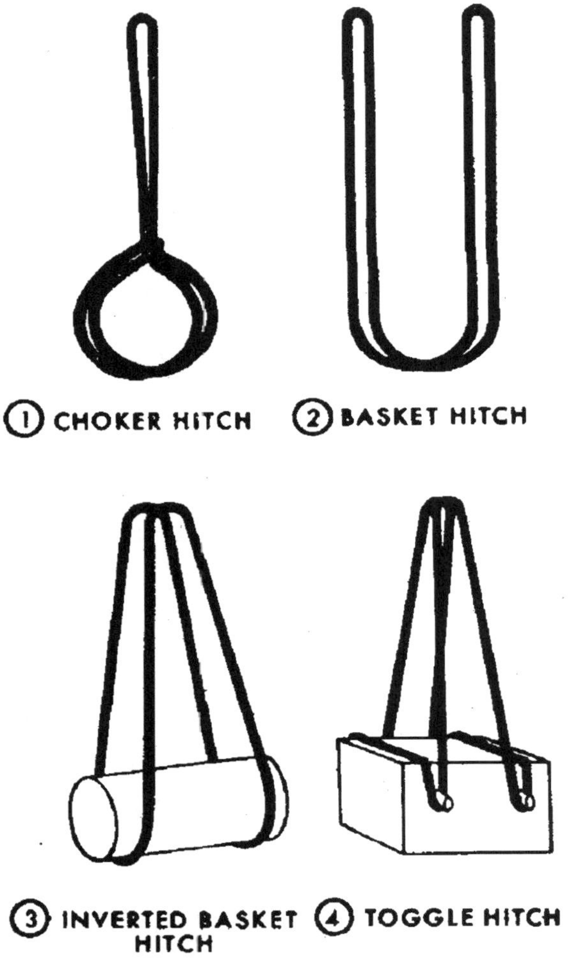

Figure 3–5. Endless slings.

b. Single Slings. A single sling can be made of wire rope, fiber rope, or chain. Each end of a single sling (fig. 3–6) is made into an eye, or has an attached hook. In some instances the ends of a wire rope are spliced into eyes around thimbles and one eye is fastened to a hook with a shackle. With this type of single sling, the shackle and hook can be removed when desired. A single sling can be used in several different ways for hoisting (fig. 3–6). It is advisable to have four single slings of wire rope available at all times. These can be used singly or in combination as may be necessary. When a single sling is used for hoisting by passing one eye through the other eye and over the hoisting hook, it is known as a choker hitch (or anchor hitch). A choker hitch will tighten down against the load when a strain is placed on the sling. If a single sling is passed under the load and both ends are hooked over the hoisting hook, it is known as a basket hitch. Single slings with two hooks which are used for lifting stone are known as stonedog hitches. Another application of a single sling is in the double anchor hitch which is used for hoisting drums or other cylindrical objects where it is necessary for the sling to tighten itself under strain and lift by friction against the sides of the cylinder.

c. Combination Slings. Single slings can be combined into bridle slings, basket slings, and choker slings (fig. 3-7) to lift virtually any type of load. Either two or four single slings can be used in a given combination. Where greater length is required, two of the single slings can be combined into a longer single sling. One of the problems in lifting heavy loads is in fastening the bottom of the sling legs to the load in such a way that the load will not be damaged. Lifting eyes are fastened to many pieces of equipment at the time it is manufactured. On large crates or boxes the sling legs may be passed under the object to form a gasket sling. A hook can be fastened to the eye on one end of each sling leg to permit easier fastening on some loads. Where the load being lifted is heavy enough or awkward enough, a four-leg sling may be required. If still greater length of sling is required, two additional slings can be used in conjunction with the four-leg sling to form a double basket.

3-8. Pallets

A problem in hoisting and moving loads sometimes occurs when the items to be lifted are packaged in small boxes and the individual boxes are not crated. In this case, it is entirely too slow to pick up each small box and move it separately. Pallets, used in combination with slings, provide an efficient method of handling such loads. Only one set of slings is required with a number of pallets (fig. 3-8). The pallets can be made up readily on the job out of 2 x 8 timbers 6 or 8 feet long, nailed to three or four heavy cross members, such as 4- x 8-inch timbers. Several pallets should be made up so that one pallet can be loaded while the pallet previously loaded is being hoisted. As each pallet is unloaded, the next return trip of the hoist takes the empty pallet back for loading.

3-9. Spreaders

Occasionally it is necessary to hoist loads that are not protected sufficiently to prevent crushing by the sling legs. In such cases, spreaders (fig. 3-9) may be used with the slings. Spread-

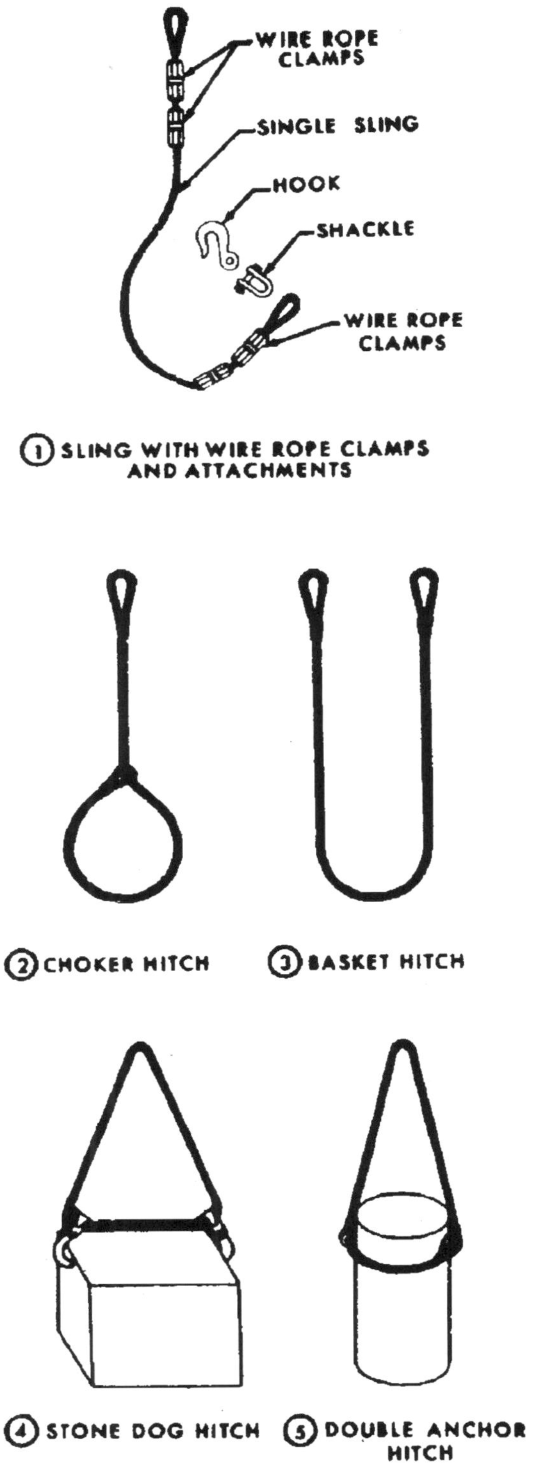

Figure 3-6. Single slings.

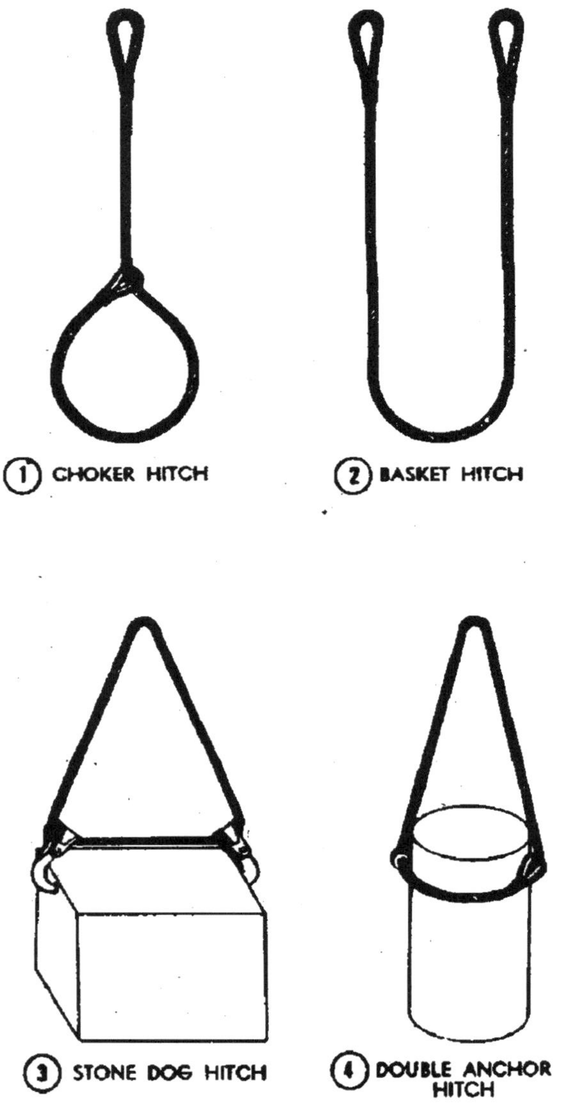

Figure 3–7. Combination slings.

ers are short bars or pipes with eyes on each end. The sling leg passes through the eye down to its connection with the load. By setting spreaders in the sling legs above the top of the load, the angle of the sling leg is changed so that crushing of the load is prevented. Changing the angle of the sling leg may increase the stress in that portion of the sling leg above the spreaders. The determining factor in computing the safe lifting capacity of the sling is the stress (or tension) in the sling leg above the spreader.

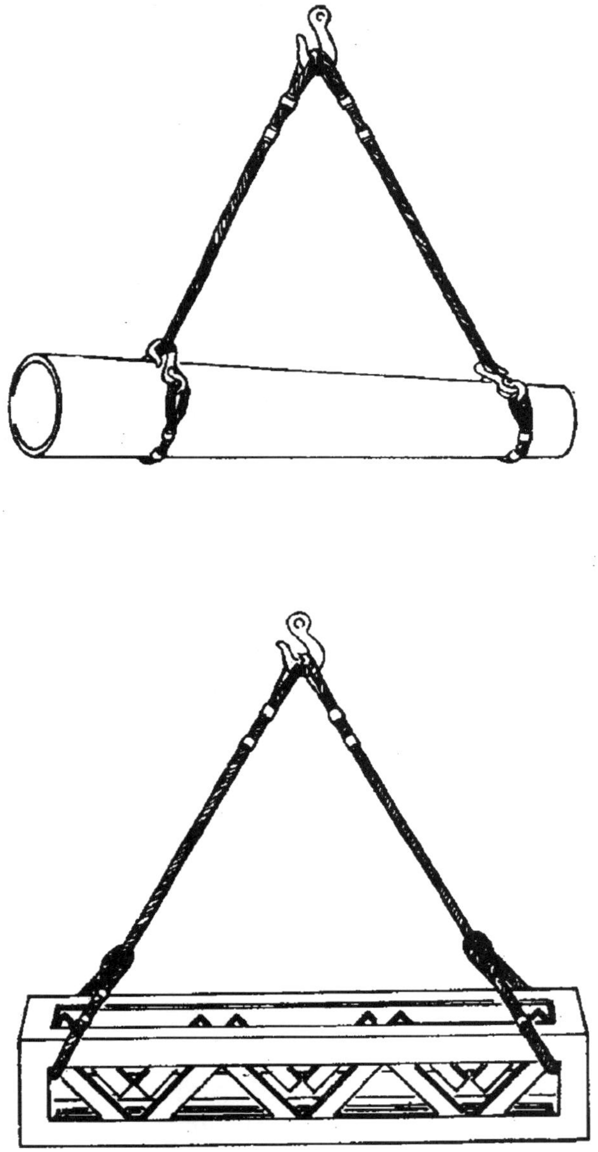

Figure 3–8. Moving loads on pallets.

3–10. Stresses

Tables 3–3 through 3–5 list the safe working loads of ropes, chains, and wire rope slings under various conditions. The angle of the legs of a sling must be considered as well as the strength of the material of which a sling is made. The lifting capacity of a sling is reduced as the angle of its legs to the horizontal is reduced (fig. 3–9) (as the legs of a sling are

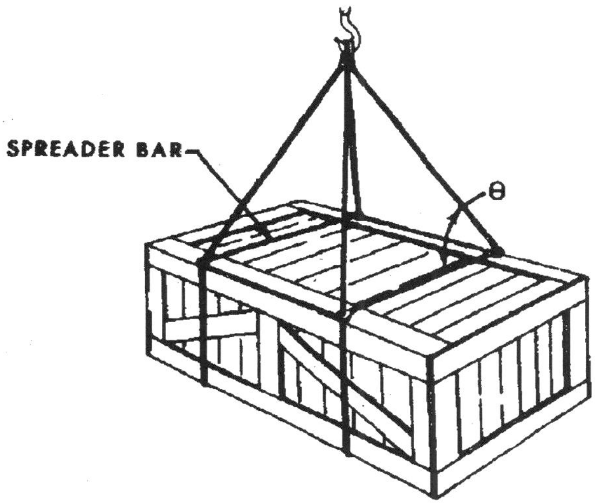

Figure 3-9. Use of spreaders in slings.

spread). Thus, the reduction of this angle of the legs of a sling increases the tension on the sling legs. In determining the proper size of sling, the tension on each leg must be determined for each load (fig. 3-10). This tension may be computed by using the following formula:

$$T = \frac{W}{N} \times \frac{L}{V}.$$

T = Tension in a single sling leg (which may be more than the weight of the load lifted)

W = Weight of the load to be lifted

N Number of slings

L Length of sling

V Vertical distance, measured from the hook to the top of the load

Note.

1. L and V must be expressed in the same unit of of measure.

2. The resulting tension will be in the same unit of measure as that of the weight of the load. Thus, if the weight of the load is in pounds the tension will be given in pounds.

Example:

Determine the tension of a single leg of a two-legged sling being used to lift a load weighing 1800 pounds. The length of a sling is 8 feet and the vertical distance is 6 feet.

Solution:

$$T = \frac{W}{N} \times \frac{L}{V}$$

$$T = \frac{1800}{2} \times \frac{8}{6} = 1200 \text{ pounds or 6 tons}$$

By knowing the amount of tension in a single leg, the appropriate size of rope, chain, or wire rope may be determined. The safe working capacity of a sling leg (keeping within the safety factors for slings) must be equal to or greater than the tension on a sling leg. If possible, the tension on each sling leg should be kept below that in the hoisting line to which the sling is attached. A particular angle formed by the sling legs with the horizontal (fig. 3-11) where the tension within each sling leg *equals* the weight of the load is called the critical angle. This angle can be approximated by the following formula:

Table 3-3. Safe Working Loads for Manila Rope Slings
(Standard, three-strand manila rope sling with a splice in each end)

Size		Single sling	Double sling			Quadruple sling		
Circumference (inches)	Diameter (inches)	Vertical lift (pounds)	60° angle (pounds)	45° angle (pounds)	30° angle (pounds)	60° angle (pounds)	45° angle (pounds)	30° angle (pounds)
¾	¼	108	187	153	108	374	306	216
1⅛	⅜	241	418	341	241	836	683	482
1½	½	475	822	672	475	1,645	1,345	950
2	⅝	791	1,370	1,119	791	2,740	2,238	1,585
2¼	¾	970	1,680	1,375	970	3,360	2,750	1,940
2¾	⅞	1,382	2,395	1,945	1,382	4,790	3,890	2,764
3	1	1,620	2,805	2,290	1,620	5,610	4,580	3,240
3½	1⅛	2,160	3,740	3,060	2,160	7,480	6,120	4,320
3¾	1¼	2,430	4,205	3,437	2,430	8,410	6,875	4,860
4½	1½	3,330	5,770	4,715	3,330	11,540	9,430	6,660
5½	1¾	4,770	8,250	6,750	4,770	16,500	13,500	9,540
6	2	5,580	9,670	7,900	5,580	19,340	15,800	11,160
7½	2½	8,366	14,500	11,850	8,366	29,000	23,700	16,732
9	3	11,520	19,950	16,300	11,520	39,900	32,600	23,040

Table 3-4. Safe Working Loads for Chain Slings
(New wrought iron chains)

Link stock diameter (inches)	Single sling	Double sling			Quadruple sling		
	Vertical lift (pounds)	60° angle (pounds)	45° angle (pounds)	30° angle (pounds)	60° angle (pounds)	45° angle (pounds)	30° angle (pounds)
3/8	2,510	4,350	3,555	2,510	8,700	7,110	5,020
7/16	3,220	5,575	4,560	3,220	11,150	9,120	6,440
1/2	4,180	7,250	5,915	4,180	14,500	11,830	8,360
9/16	5,420	9,375	7,670	5,420	18,750	15,340	10,840
5/8	6,460	11,175	9,150	6,460	22,350	18,300	12,920
3/4	9,160	15,850	12,950	9,160	31,700	25,900	18,320
7/8	13,020	22,550	18,410	13,020	45,100	36,820	26,000
1	17,300	29,900	24,450	17,300	59,800	48,900	34,600
1 1/8	21,550	37,350	30,550	21,550	74,700	61,100	43,100
1 1/4	26,600	46,050	37,600	26,600	92,100	75,200	53,200
1 3/8	32,200	55,750	45,600	32,200	111,500	91,200	64,400
1 1/2	38,300	66,400	54,250	38,300	132,800	108,500	76,600
1 5/8	44,600	77,200	63,050	44,600	154,400	126,100	89,200
1 3/4	51,300	88,750	72,500	51,300	177,500	145,000	102,600
1 7/8	58,700	101,500	83,000	58,700	203,000	166,000	117,400
2	66,200	114,500	93,500	66,200	229,000	187,000	132,400

Table 3-5. Safe Working Loads for Wire Rope Slings
(New improved plow steel wire rope)

Diameter (inches)	Single sling	Double sling			Quadruple sling		
	Vertical lift (pounds)	60° angle (pounds)	45° angle (pounds)	30° angle (pounds)	60° angle (pounds)	45° angle (pounds)	30° angle (pounds)
1/4	1,096	1,899	1,552	1,096	3,798	3,105	2,192
5/16	1,690	2,925	2,390	1,690	5,850	4,780	3,380
3/8	2,460	4,260	3,485	2,460	8,520	6,970	4,920
7/16	3,560	6,170	5,040	3,560	12,340	10,080	7,120
1/2	4,320	7,475	6,105	4,320	14,950	12,210	8,640
9/16	5,460	9,450	7,725	5,460	18,900	15,450	10,920
5/8	6,650	11,500	9,400	6,650	23,000	18,800	13,300
3/4	9,480	16,400	13,400	9,480	32,800	26,800	18,960
7/8	12,900	22,350	18,250	12,900	44,700	36,500	25,800
1	16,800	29,100	23,750	16,800	58,200	47,500	33,600
1 1/8	21,200	36,700	30,000	21,200	73,400	60,000	42,400
1 1/4	26,000	45,000	36,800	26,000	90,000	73,600	52,000
1 3/8	32,000	55,400	45,250	32,000	110,800	90,500	64,000
1 1/2	37,000	64,000	52,340	37,000	128,000	104,700	74,000
1 5/8	41,800	72,400	59,200	41,800	144,800	118,400	83,600
1 3/4	49,800	86,250	70,500	49,800	172,500	141,000	99,600
2	62,300	107,600	88,050	62,300	215,200	176,100	124,600
2 1/4	82,900	143,500	117,400	82,900	287,000	234,800	165,800
2 1/2	101,800	176,250	144,000	101,800	352,500	288,000	203,600
2 3/4	122,500	212,000	173,500	122,500	424,000	347,000	245,000

Critical angle $= \frac{60}{N}$ (N = number of sling legs)

It is desirable to stay above the critical angle when using slings.

3-11. Inspection and Cushioning

Slings should be inspected periodically and condemned when they are no longer safe. The usual deterioration check for fiber ropes, wire ropes, chains, and hooks, must be made when they are used in slings. In addition to the usual precautions, wire ropes used in slings are declared unsafe if 4 percent or more of the wires are broken. Objects to be lifted must be padded with wood blocks, heavy fabric, old rubber tires, or other cushioning material to protect the legs of slings from being damaged.

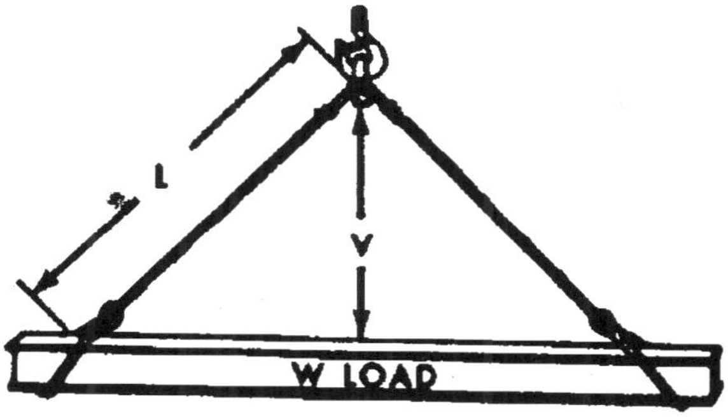

Tension in a sling leg

$$T = \frac{W_L}{N} \times \frac{L}{V}$$

T = Tension in a single leg

W_L = Weight of load

N = Number of sling legs

L = Length of sling leg

V = Vertical distance of sling

Figure 3–10. Computing tension in a sling.

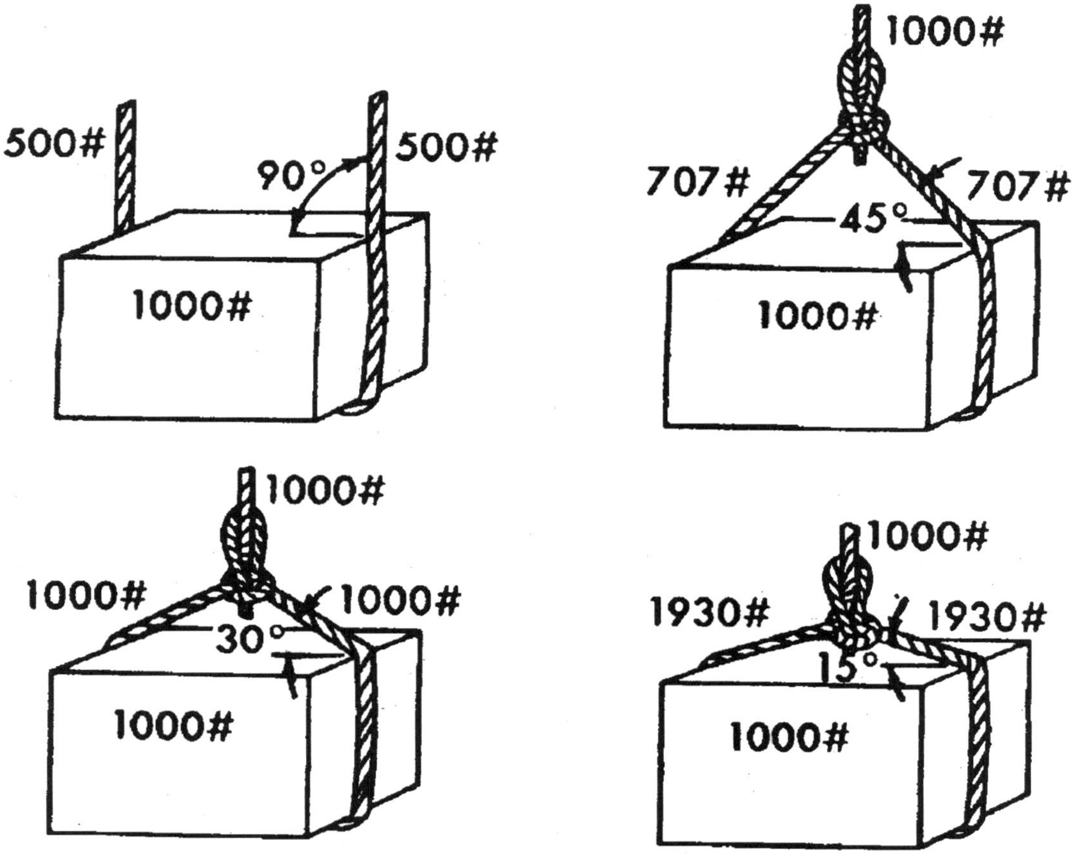

(1) Critical Angle— The sling angle that exists when the tension in the sling leg <u>equals</u> the weight of the load.

(2) Critical Angle Formula:

CA = 60°

N = Number of sling legs

Figure 3–11. Sling angles.

Section III. MECHANICAL ADVANTAGE

3–12. Introduction

A force is a push or pull. The push or pull a human can exert depends on his weight and strength. In order to move any load heavier than the maximum amount a man can move, a machine must be used to multiply the force exerted into a force capable of moving the load. The machine used may be a lever, a screw, or a tackle system. The same principle applies to all of these. If a machine is used which exerts a force 10 times greater than the force applied to it, the machine has multiplied the force input by 10. The mechanical advantage of a machine is the amount by which the machine multiplies the force applied to it in order to lift or move a load. For example, if a downward push of 10 pounds on the left end of a lever will cause the right end of the lever to raise a load weighing 100 pounds, the lever is said to have a mechanical advantage of 10.

3–13. Blocks and Tackle

A block (A, fig. 3–12) consists essentially of a wood or metal frame containing one or more rotating pulleys called sheaves. A tackle is an assembly of ropes and blocks used to multiply force. (B, fig. 3–12) The number of times the force is multiplied is the mechanical advantage of the tackle. To make up a tackle system, the blocks to be used are laid out and the rope is reeved (threaded) through the blocks. A simple tackle is one or more blocks reeved with a single rope. Compound tackle is two or more blocks reeved with more than one rope. Every tackle system contains a fixed block attached to some solid support, and may have a traveling block attached to the load. The single rope leaving the tackle system is called the fall line. The pulling force is applied to the fall line, which may be led through a leading block. This is an additional block used to change the direction of pull.

a. Blocks. Blocks are used to reverse the direction of rope in tackle. Blocks (fig. 3–13) take their names from the purpose for which they are used, the places they occupy, or from a particular shape or type of construction. According the the number of sheaves, blocks are designated as single, double, or triple. A snatch

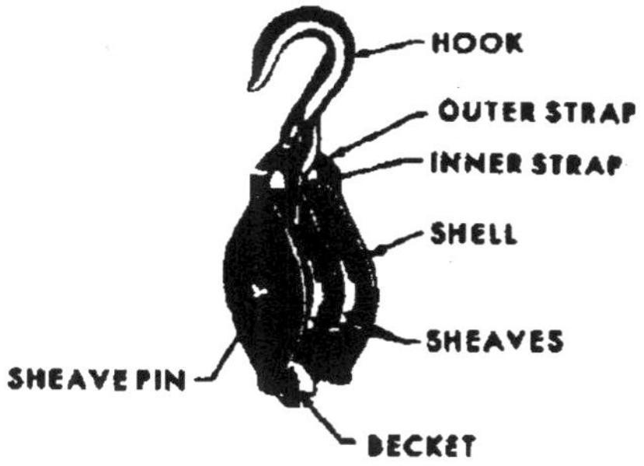

A. Components of a double block.

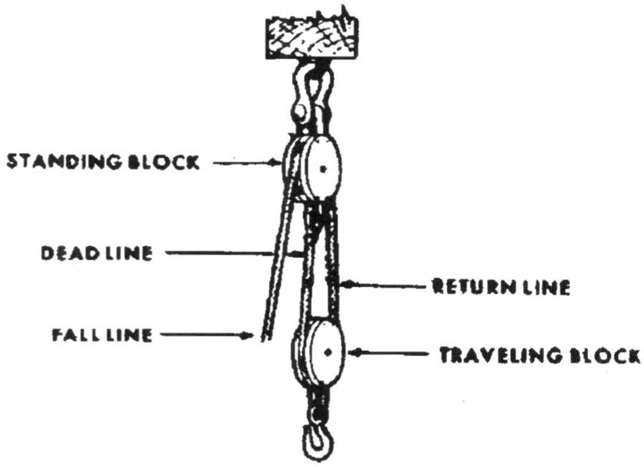

Figure 3–12. Double block and a tackle system.

block is a single sheave block made so that the shell opens on one side at the base of the hook to permit a rope to be slipped over the sheave without threading the end of it through the block. Snatch blocks ordinarily are used where it is necessary to change the direction of the pull on the line. A traveling block is a block attached to the load which is being lifted and moves as the load is lifted. A standing block is a block that is fixed to a stationary object.

(1) *Leading blocks.* Blocks used in the tackle to change the direction of the pull without affecting the mechanical advantage of the system are called leading blocks (fig. 3–14). In some tackle systems the fall line leads off

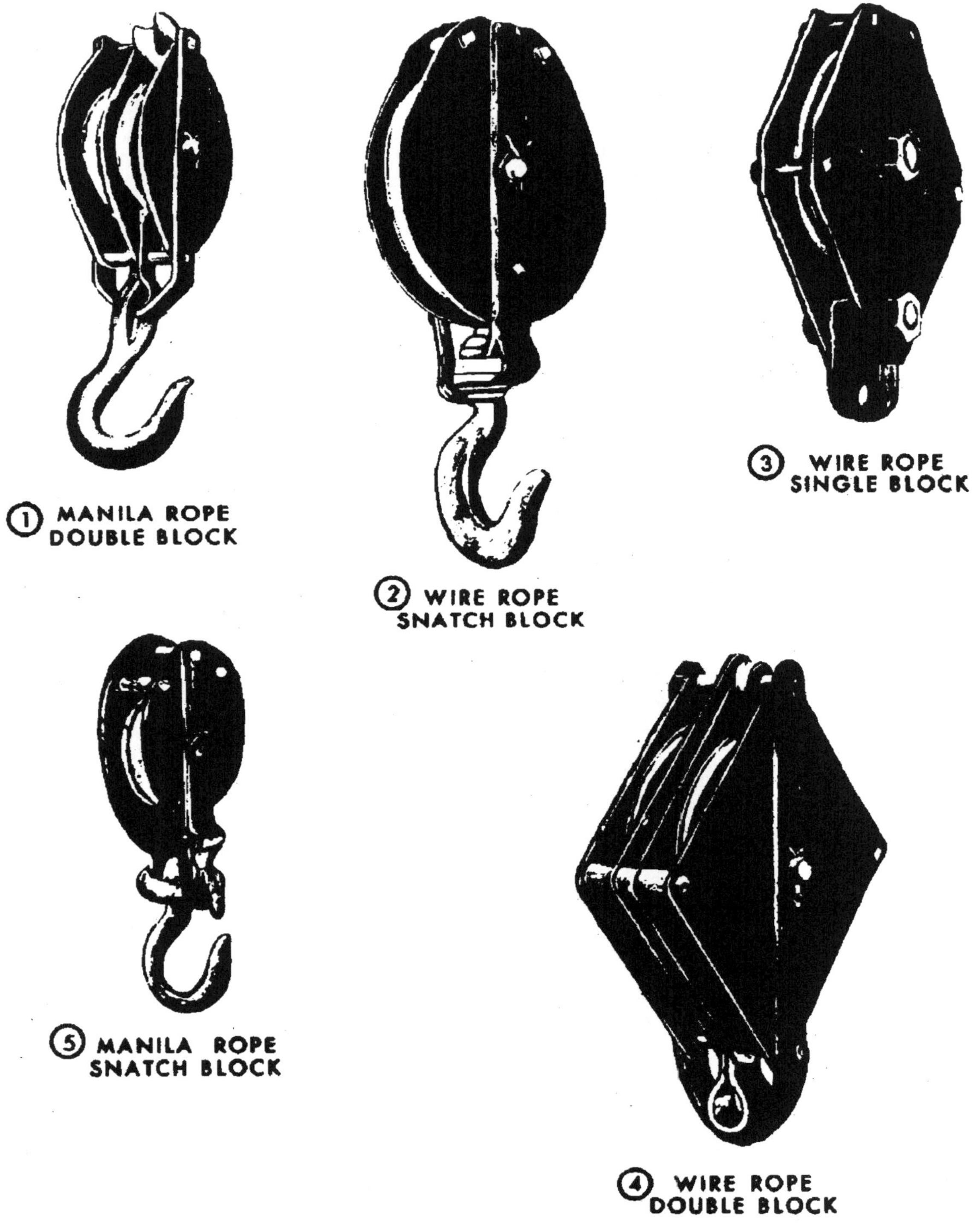

Figure 3–13. Types of blocks.

the last block in a direction which makes it difficult to apply the motive force required. A leading block is used to correct this. Ordinarily a snatch block is used as the leading block. This block can be placed at any convenient position. The fall line from the tackle system is led through the leading block to the line of most direct action.

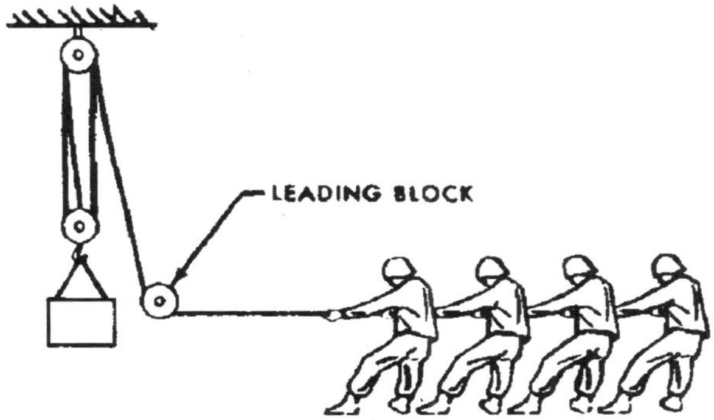

Figure 3–14. Use of leading block.

(2) *Reeving blocks.* Blocks are laid out for reeving on a clean and level surface other than the ground to avoid getting dirt into the operating parts. Figure 3–15 shows the reeving of single and double blocks. In reeving triple blocks (fig. 3–16), it is imperative that the hoisting strain be put at the center of the blocks to prevent them from being inclined under the strain. If the blocks do incline, the rope will drag across the edges of the sheaves and the shell of the block and cut the fibers. The blocks are placed so that the sheaves in one block are at right angles to the sheaves in the other block. The coil of rope may be laid beside either block. The running end is passed over the center sheave of one block and back to the bottom sheave of the other block. It is then passed over one of the side sheaves of the first block. In selecting which side sheave over which to pass the rope, one must remember that the rope should not cross the rope leading away from the center sheave of the first block. The rope is then led over the top sheave of the second block and back to the remaining side sheave of the first block. From this point, the rope is led to the center sheave of the second block and back to the becket of the first block. The rope should be reeved through the blocks so that no part of the rope chafes another part of the rope.

(3) *Twisting.* Blocks should be reeved in a manner that prevents twisting. After the blocks are reeved, the rope should be pulled back and forth through the blocks several times to allow the rope to adjust to the blocks. This reduces the tendency of the tackle to twist under a load. When the ropes in a tackle system become twisted, there is an increase in friction and chafing of the ropes, as well as a possibility of jamming the blocks. When the hook of the standing block is fastened to the supporting member, the hook should be turned so that the fall line leads directly to the leading block or to the source of motive power. It is very difficult to prevent twisting of a traveling block. It is particularly important when the tackle is being used for a long pull along the ground, such as in dragging logs or timbers. One of the simplest antitwisting devices for such a tackle is a short iron rod or piece of pipe lashed to the traveling block (fig. 3–17). The antitwisting rod or pipe may be lashed to the shell of the block with two or three turns of rope. If it is lashed to the becket of the block, the rod or pipe should pass between the ropes without chafing them as the tackle is hauled in.

b. Simple Tackle Systems. A simple tackle system is one using one rope and one or more blocks. To determine the mechanical advantage of a simple system (fig. 3–18), count the number of lines supporting the load (or the traveling block). In counting, the fall line is included if it leads out of a traveling block. In a simple tackle system the mechanical advantage always will be the same as the number of lines supporting the load. As an alternate method, the mechanical advantage can be determined by tracing the forces through the system. Thus, begin with a unit force applied to the fall line. Assume that the tension in a single rope is the same throughout and therefore the same force will exist in each line. Total all the forces acting on the load or traveling block. The ratio of the resulting total force acting on the load or traveling block to the original unit force exerted on the fall line is the theoretical mechanical advantage of the simple system.

Examples:

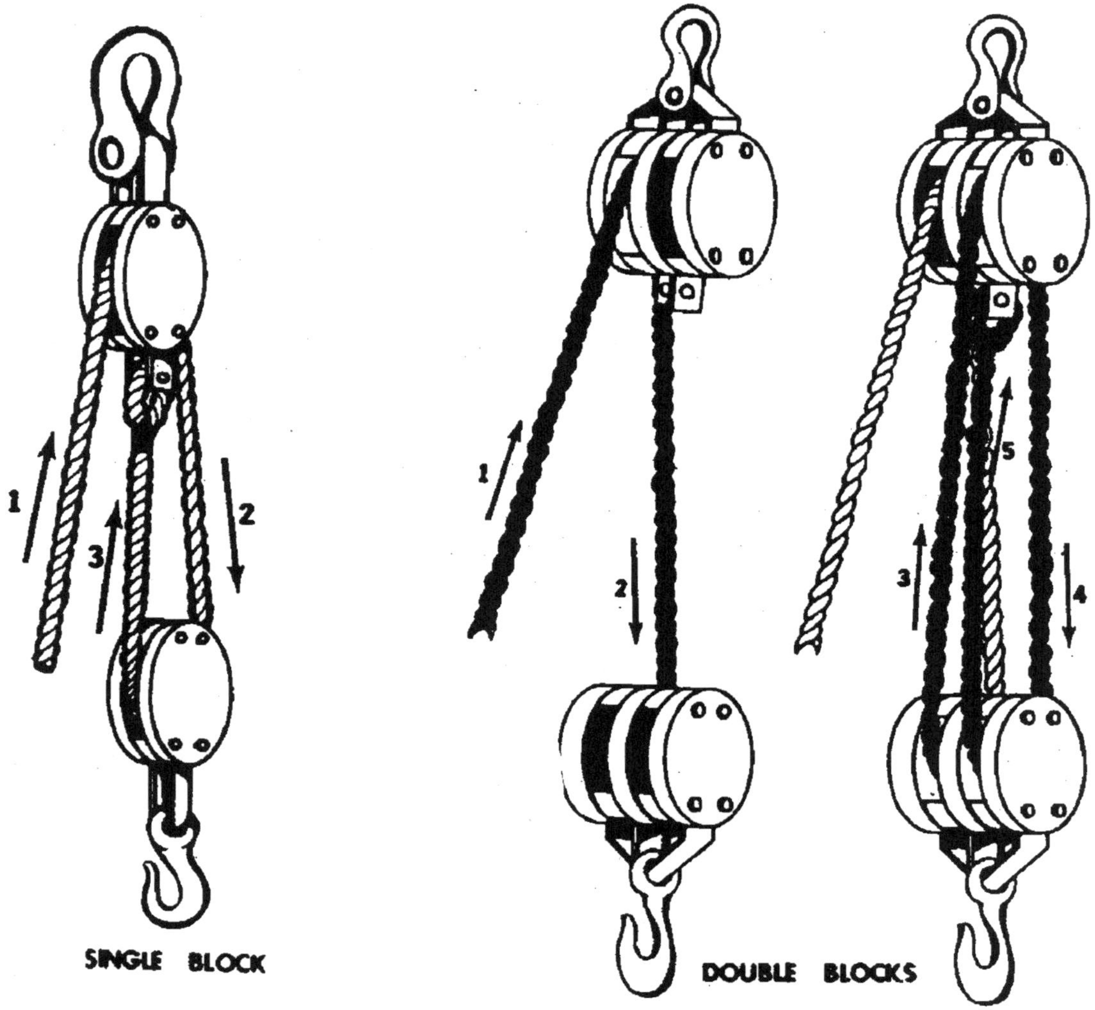

Figure 3–15. Reeving single and double blocks.

Method I—Counting Supporting Lines (①, fig. 3–19). There are three lines supporting the traveling block, so the theoretical mechanical advantage is 3:1.

Method II—Unit Force (②, fig. 3–19). Assuming the tension on a single rope is the same throughout its length, a unit force of 1 on the fall line results in a total of 3 unit forces acting on the traveling block. The ratio of the resulting force of 3 on the traveling block to the unit force of 1 on the fall line gives a theoretical mechanical advantage of 3:1.

c. *Compound Tackle Systems.* A compound tackle system (fig. 3–20), is one using more than one rope with two or more blocks. Compound systems are made up of two or more simple systems. The fall line from one simple system is fastened to a hook on the traveling block of another simple system, which may include one or more blocks. In compound systems the mechanical advantage can best be determined by using the unit force method. Begin with a unit force applied to the fall line. Assume that the tension in a single rope is the same throughout and therefore the same force will exist in each line. Total all the forces acting on the traveling block and transfer this

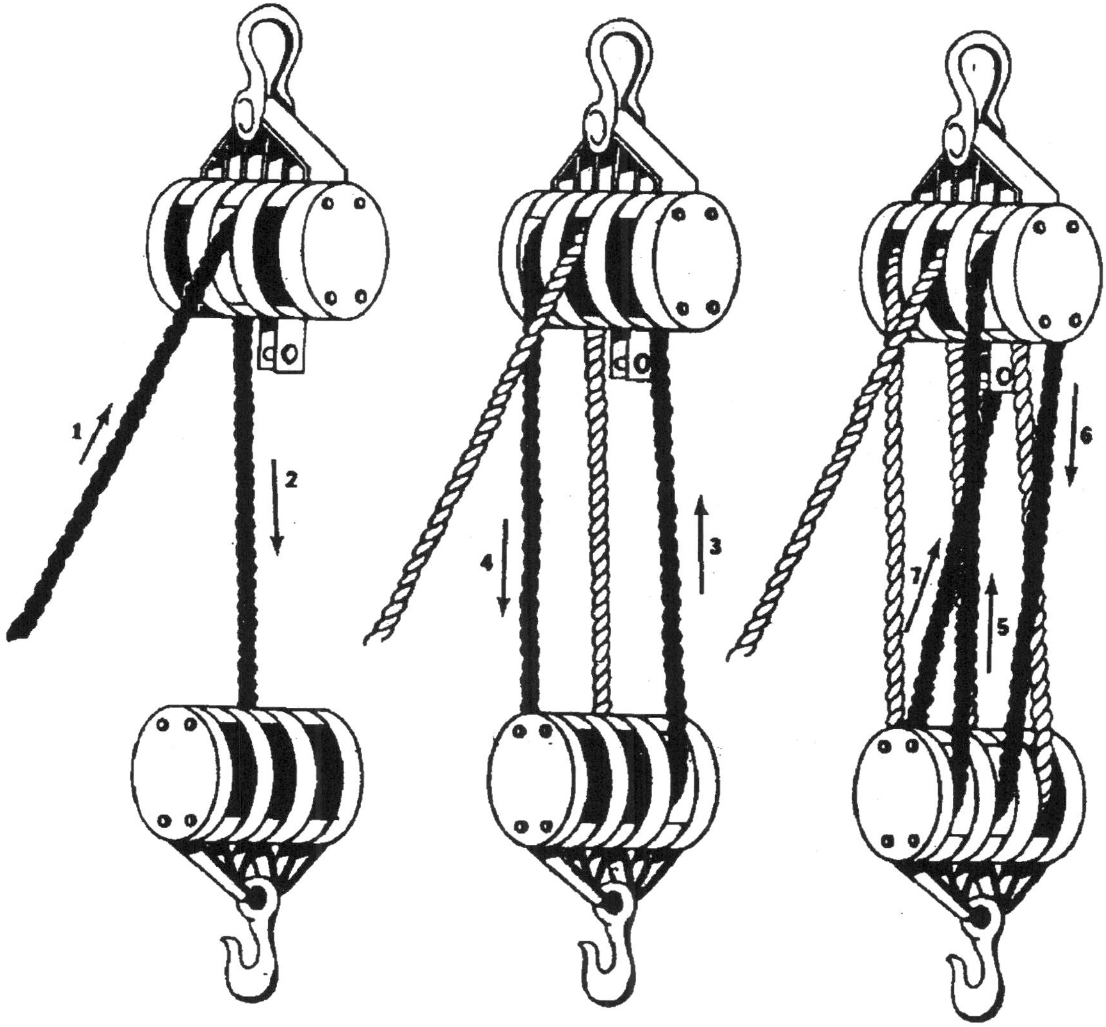

Figure 3–16. Reeving triple blocks.

Figure 3–17. Antitwisting rod or pipe.

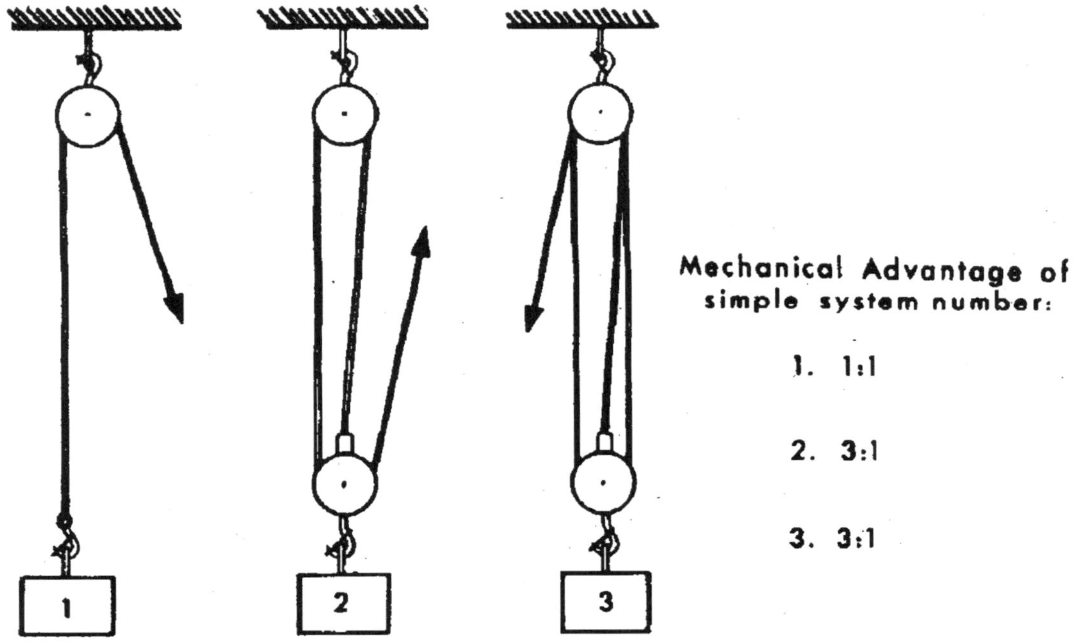

Figure 3–18. Simple tackle systems.

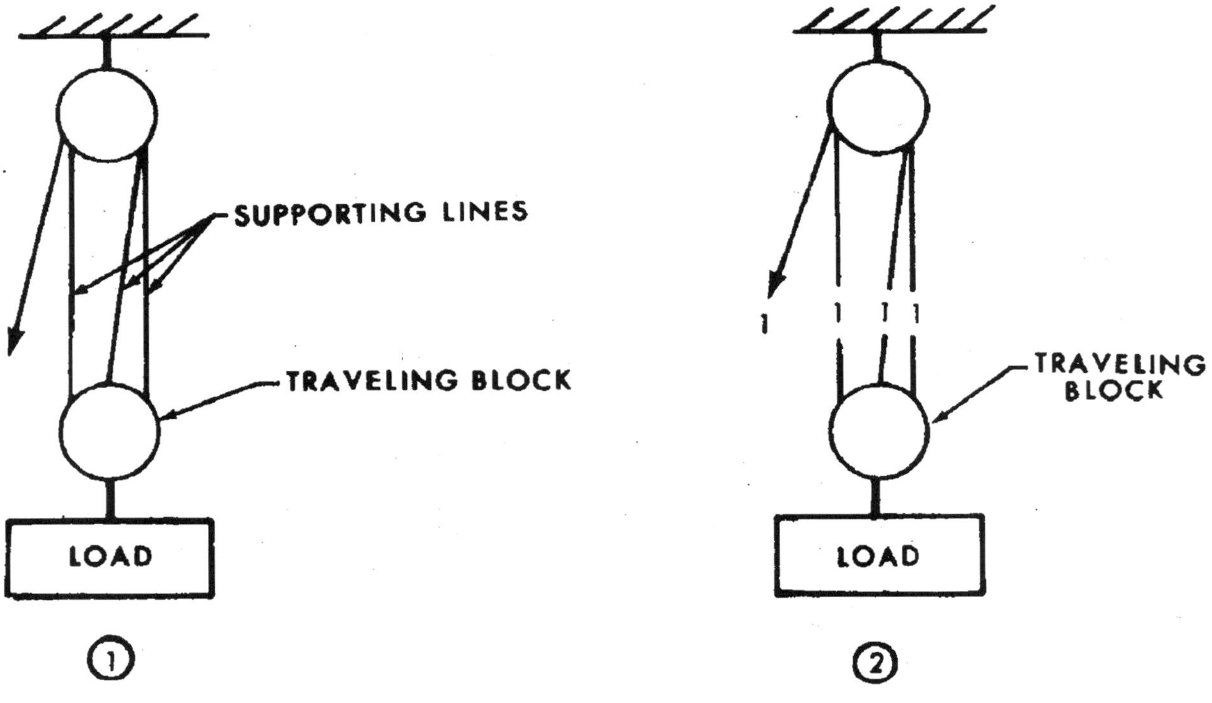

Figure 3–19. Determining ratio of a simple tackle.

force into the next simple system. The ratio of the resulting total force acting on the load or traveling block to the original unit force exerted on the fall line is the theoretical mechanical advantage of the compound system. Another method, simpler, but less accurate in some cases, is by determining the mechanical advantage of each simple system in the compound system and multiplying these together to obtain the total mechanical advantage.

Examples:

Method I—Unit Force (①, fig. 3–21).

As in method II of *simple* tackle systems, a unit force of 1 on the fall line results in 4 unit forces acting on the traveling block of tackle system A. Transferring the unit force of 4 into the fall line of simple system B results in a total of 16 unit forces (4 lines with 4 units of force in each) acting on the traveling block of tackle system B. The ratio of 16 unit forces on the traveling block carrying the load to a 1 unit force on the fall line gives a theoretical mechanical advantage of 16:1.

Method II—Multiplying Mechanical Advantages of Simple Systems (②, fig. 3–21). The number of lines supporting the traveling blocks in both systems number A and B is equal to 4. The mechanical advantage of each simple system is therefore equal to 4:1. The mechanical advantage of the compound system is then determined by multiplying together the mechanical advantage of each simple system for a resulting mechanical advantage of 16:1.

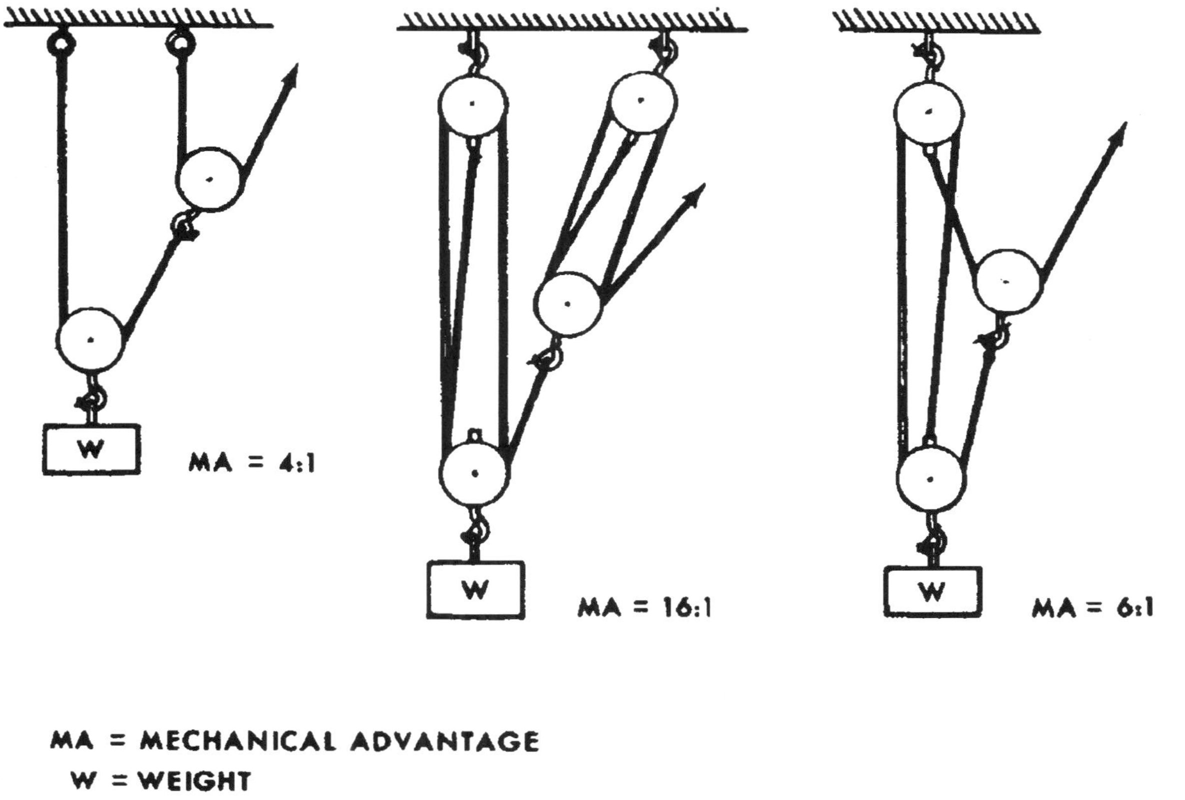

Figure 3–20. Compound tackle systems.

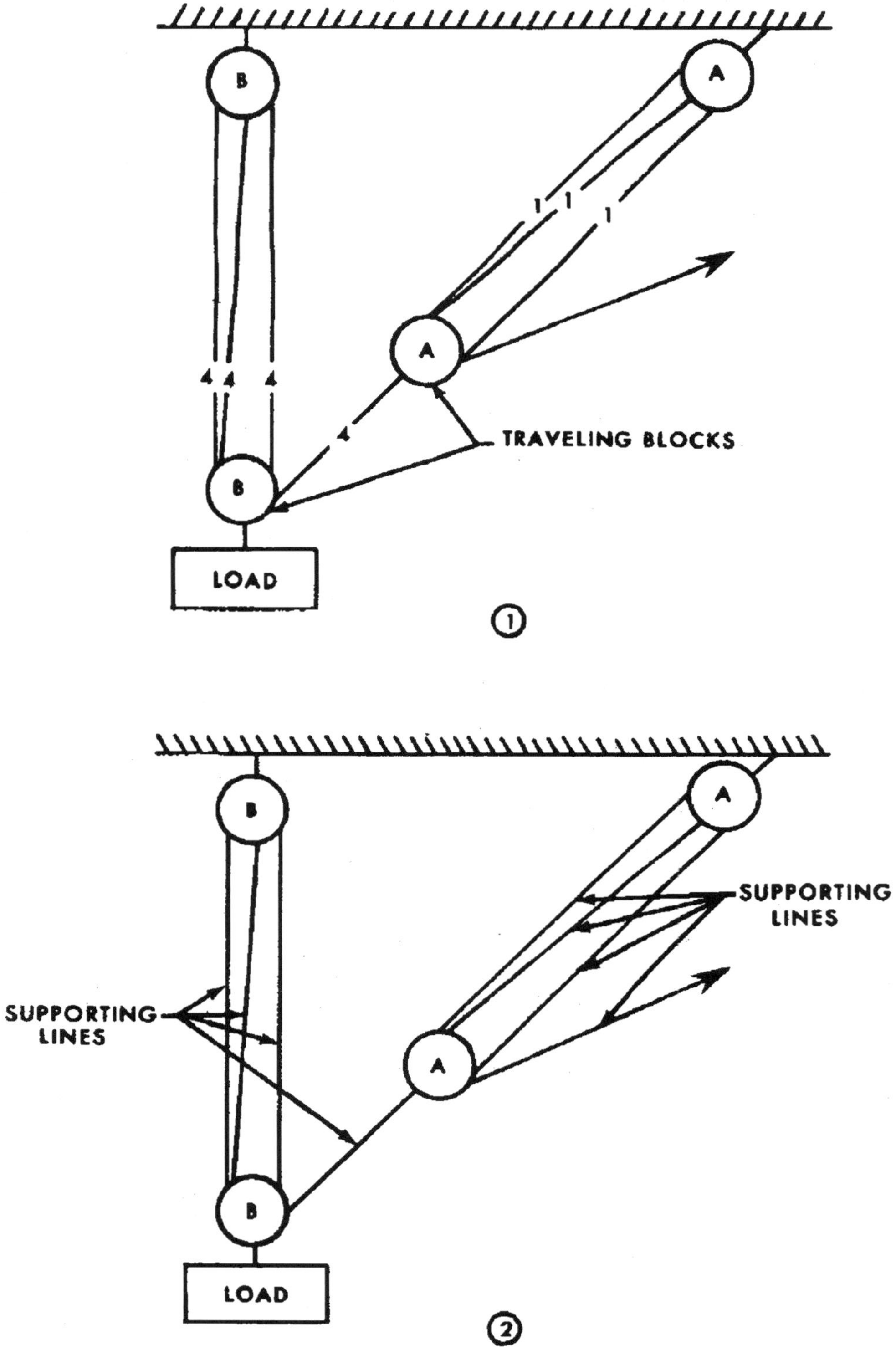

Figure 3–21. Determining ratio of a compound tackle system.

d. Friction. There is a loss in any tackle system due to the friction created by the sheave rolling on the pin, the ropes rubbing together, and the rope rubbing against the sheave. This friction reduces the total lifting power. Therefore, the force exerted on the fall line must be increased by some amount to overcome the friction of the system in order to lift the load. Each sheave in the tackle system can be expected to create a resistance equal to approximately 10 percent of the weight of the load. For example, a load weighing 5,000 pounds is lifted by a tackle system which has a mechanical advantage of 4:1. The rope travels over 4 sheaves which produce a resistance of 40 percent of 5,000 pounds or 2,000 pounds (5,000 x .40). The actual pull that would be required on the fall line of the tackle system is equal to the summation of the weight of the load *and* the friction in the tackle system divided by the theoretical mechanical advantage of the tackle system. In this example, the actual pull required on the fall line would be equal to the sum of 5,000 lbs (Load) and 2,000 lbs (Friction) divided by 4 (Mechanical Advantage) or 1,750 lbs. There are other types of resistance which may have to be considered in addition to tackle resistance. A thorough discussion of resistance is presented in FM 20–22, Vehicle Recovery Operations.

e. Source of Power. In all cases where manpower is used for hoisting, the system must be arranged to consider the most satisfactory method of utilizing that source of power. More men can pull on a single horizontal line along the ground than on a single vertical line. On a vertical pull, men of average weight can pull approximately 100 pounds per man, and on a horizontal pull approximately 60 pounds per man. If the force required on the fall line is 300 pounds or less, the fall line can lead directly down from the upper block of a tackle vertical line. If 300 pounds times the mechanical advantage of the system is not enough to lift a given load, the tackle must be re-rigged to increase the mechanical advantage, or the fall line must be led through a leading block to provide a horizontal pull. This will permit more men to pull on the line. Similarly, if a heavy load is to be lifted and the fall line is led through a leading block to a winch mounted on a vehicle, the full power available at the winch is multiplied by the mechanical advantage of the system.

Section IV. METHODS

3–14. Chain Hoists

Chain hoists (fig. 3–22) provide a convenient and efficient method for hoisting by hand under particular circumstances. The chief advantages of chain hoists are that the load can remain stationary without requiring attention, and that the hoist can be operated by one man to raise loads weighing several tons. The slow lifting travel of a chain hoist permits small movements, accurate adjustments of height, and gentle handling of loads. A ratched handle pull hoist (fig. 3–23) is used for short horizontal pulls on heavy objects. Chain hoists differ widely in their mechanical advantage, depending upon their rated capacity which may vary from 5 to 250.

a. Types. There are three general types of chain hoists for vertical operation—the differential chain hoist, the spur gear hoist, and the screw gear hoist. The spur gear hoist is the most satisfactory for ordinary operation where a minimum number of men are available to operate the hoist and the hoist is to be used frequently. This type of hoist is about 85 percent efficient. The screw gear hoist is about 50 percent efficient and is satisfactory where less frequent use of the hoist is involved. The differential hoist is only about 35 percent efficient, but is satisfactory for occasional use and light loads.

b. Safety. Chain hoists are usually stamped with their load capacities on the shell of the upper block. The rated load capacity will run from one-half of a ton upward. Ordinarily, chain hoists are constructed with their lower hook as the weakest part of the assembly. This is done as a precaution, so that the lower hook will be overloaded before the chain hoist is overloaded. The lower hook will start to spread under overload, indicating to the operator that he is approaching the overload point of the chain hoist. Under ordinary circumstances the pull exerted on a chain hoist by one or two men will not overload the hoist. Chain hoists should

Figure 3–22. Chain hoists.

be inspected at frequent intervals. Any evidence of spreading of the hook or excessive wear is sufficient cause to require replacement of the hook. If the links of the chain are distorted, it indicates that the chain hoist has been heavily overloaded and is probably unsafe for further use. Under such circumstances the chain hoist should be condemned.

3–15. Winches

Vehicular-mounted winches and engine-driven winches are used with tackles for hoisting (fig. 3–24). There are two points to consider when placing a power-driven winch to operate hoisting equipment; first, the angle with the ground which the hoisting line makes at the drum of the hoist, and second, the fleet angle (fig. 3–25) of the hoisting line winding on the drum. The

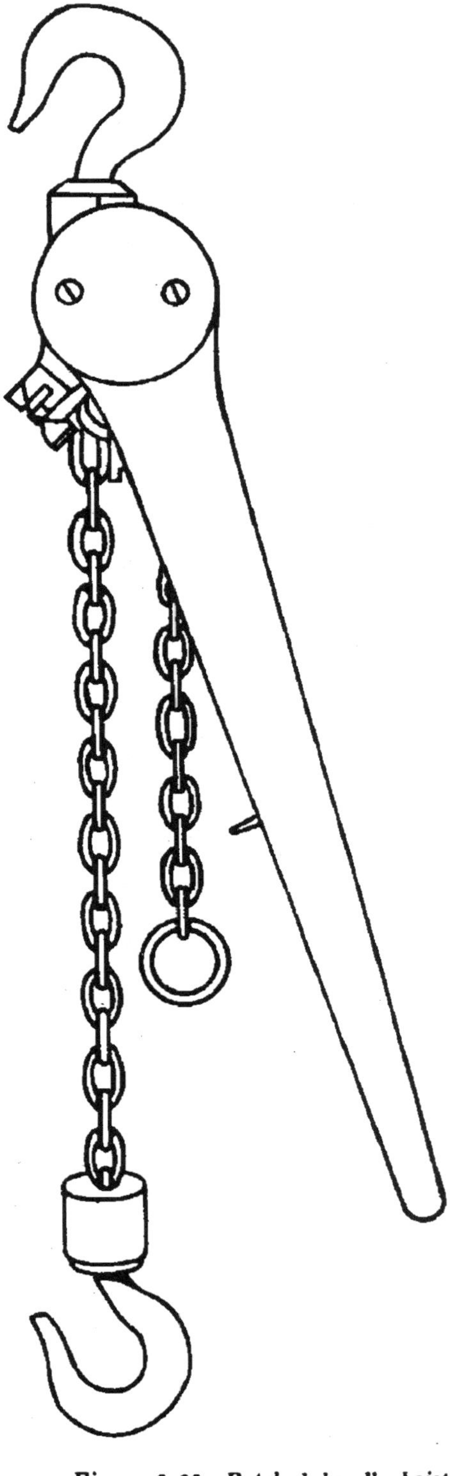

Figure 3–23. Ratched handle hoist.

distance from the drum to the first sheave of the system is the controlling factor in the fleet angle. When using vehicular-mounted winches, the vehicle should be placed in a position which permits the operator to watch the load being hoisted. A winch is most effective when the pull is exerted on the bare drum of the winch. When a winch is rated at a capacity, that rating applies only as the first layer of cable is wound onto the drum. The winch capacity is reduced as each layer of cable is wound onto the drum because of the change in leverage resulting from the increased diameter of the drum. The capacity of the winch may be reduced by as much as 50 percent when the last layer is being wound onto the drum.

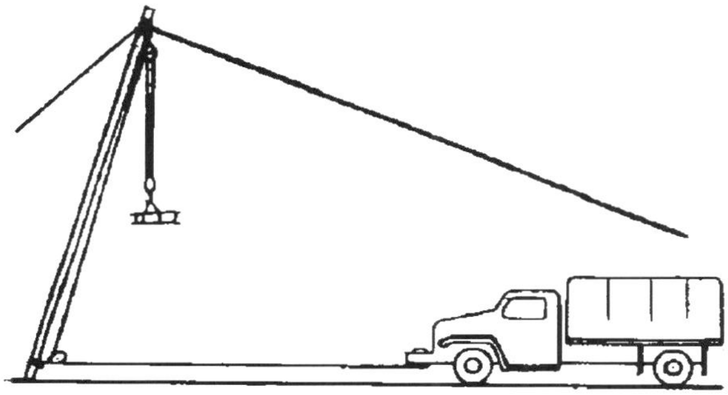

Figure 3-24. Using a vehicular winch for hoisting.

a. Ground Angle. If the hoisting line leaves the drum at an angle upward from the ground, the resulting pull on the winch will tend to lift it clear of the ground. In this case a leading block must be placed in the system at some distance from the drum to change the direction of the hoisting line to a horizontal or downward pull. The hoisting line should be overwound or underwound on the drum as may be necessary to avoid a reverse bend.

b. Fleet Angle. The drum of the winch is placed so that a line from the last block passing through the center of the drum is at right angles to the axis of the drum. The angle between this line and the hoisting line as it winds on the drum is called the fleet angle (fig. 3-25). As the hoisting line is wound in on the drum, it moves from one flange to the other, so that the fleet angle changes during the hoisting process. The fleet angle should not be per-

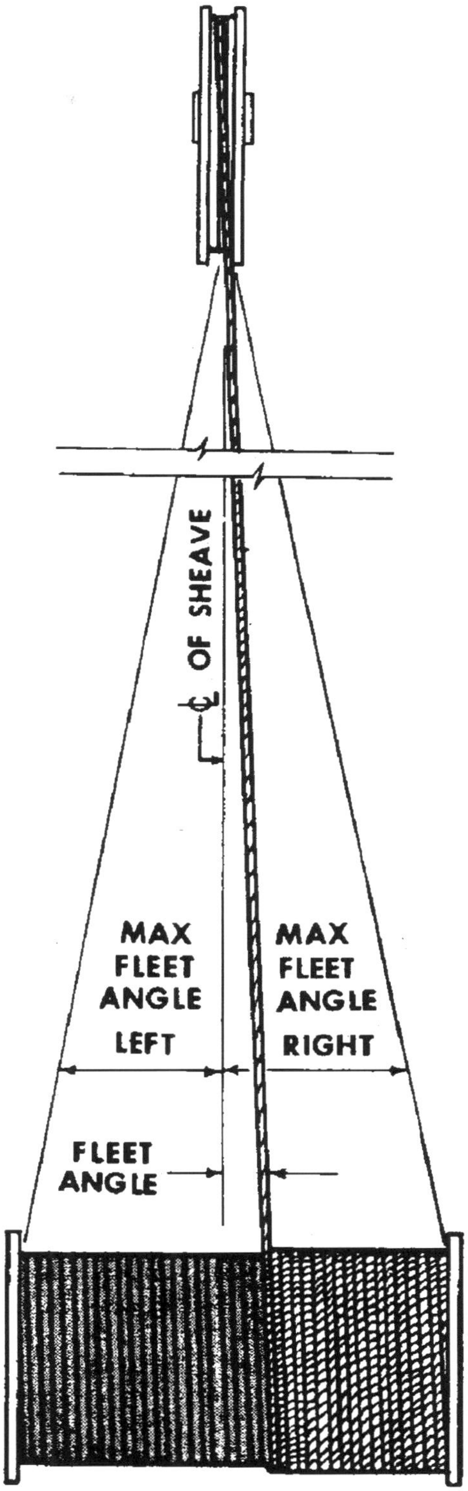

Figure 3-25. Fleet angle.

mitted to exceed 2° and should be kept below this if possible. A 1½-degree maximum angle is satisfactory and will be obtained if the distance from the drum to the first sheave is 40 inches for each inch from the center of the drum to the flange. The wider the drum of the hoist the greater the lead distance must be in placing the winch.

3–16. Expedients

In the absence of mechanical power or an appropriate tackle, it may be necessary to use makeshift equipment for hoisting or pulling. A Spanish windlass can be used to move a load along the ground, or the horizontal pull from the windlass can be directed through blocks to provide a vertical pull on a load. In making a Spanish windlass, a rope is fastened between the load to be moved and an anchorage some distance away. A short spar is placed vertically beside this rope (fig. 3–26), approximately halfway between the anchorage and the load. This spar may be a pipe or a pole, but in either case should have as large a diameter as possible. A loop is made in the rope and wrapped partly around the spar. The end of a horizontal rod is inserted through this loop. The horizontal rod should be a stout pipe or bar long enough to provide leverage. It is used as a lever to turn the vertical spar. As the vertical spar turns, the rope is wound around it which shortens the line and pulls on the load. The rope leaving the vertical spar should be as near the same level as possible on both sides to prevent the spar from tipping over.

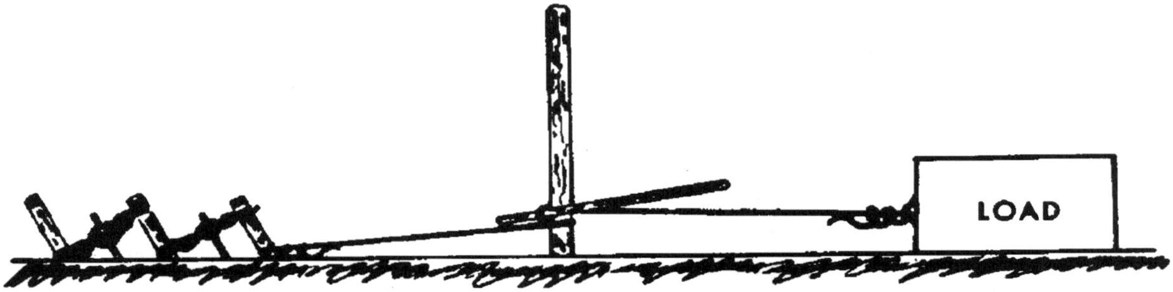

Figure 3–26. Spanish windlass.

CHAPTER 4

ANCHORAGES AND GUYLINES

Section I. ANCHORS

4–1. Introduction

When heavy loads are handled with a tackle, it is necessary to have some means of anchorage. Many expedient rigging installations are supported by combining the use of guylines and some type of anchorage system. Anchorage systems may be either natural or manmade. The type of anchorage to be used will depend on the time and material available, and on the holding power required. Whenever possible, natural anchorages should be utilized so that time, effort, and material may be conserved. The ideal anchorage system must be of sufficient strength to support the breaking strength of the attached line. Lines should always be fastened to anchorages at a point as near to the ground as possible. The principle factor in the strength of most anchorage systems is the area bearing against the ground.

4–2. Natural Anchors

Trees, stumps, or rocks can serve as natural anchorages for rapid work in the field. Always attach lines near the ground level on trees or stumps (fig. 4–1). Avoid dead or rotten trees or stumps as an anchorage because they are likely to snap suddenly when a strain is placed on the line. It is always advisable to lash the first tree or stump to a second one, to provide added support. A transom (fig. 4–2) can be placed between two trees to provide stronger anchorage than a single tree. When using rocks (fig. 4–3) as natural anchorages, examine the rocks carefully to be sure that they are large enough and

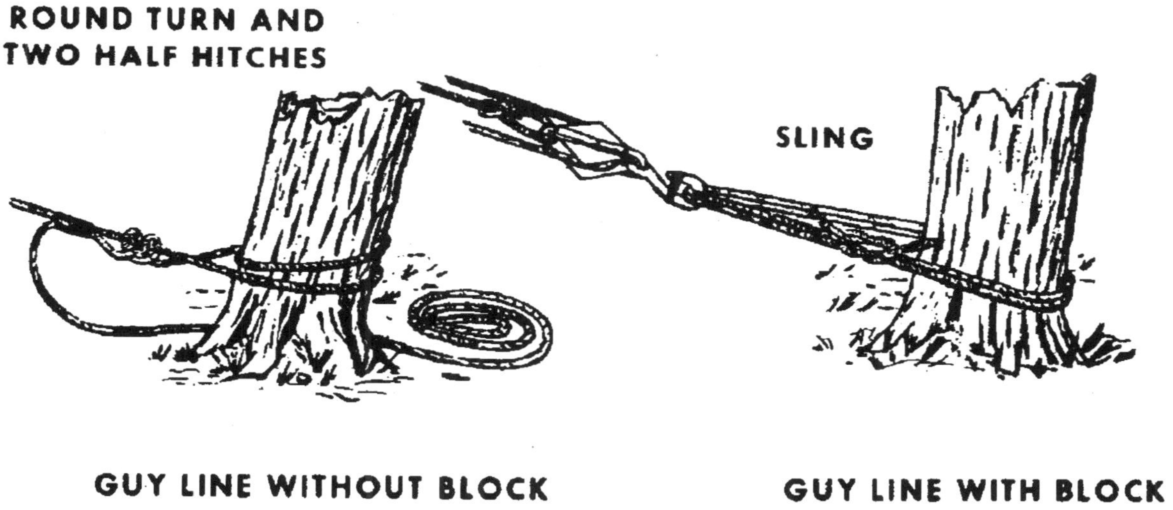

Figure 4 1. Natural anchorage—tree.

firmly embedded in the ground. An outcropping of rock or a heavy boulder buried partially in the ground will serve as a satisfactory anchor.

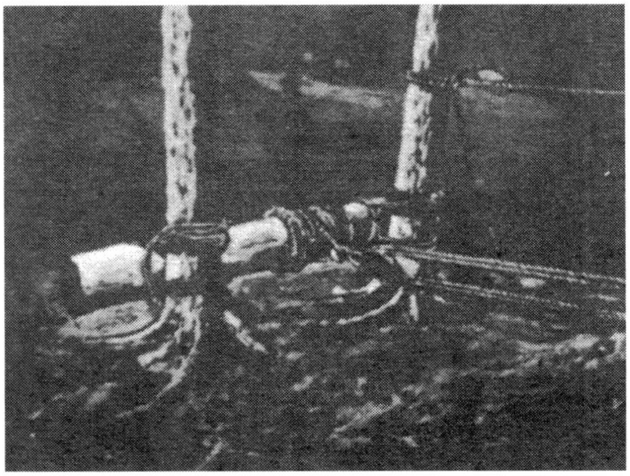

Figure 4–2. Natural anchorage—trees and transom.

Figure 4–3. Natural anchorage—rock.

4–3. Manmade Anchors

Manmade anchors must be constructed when natural anchors are not available.

a. Rock Anchor. The rock anchor (fig. 4–4) has an eye on one end and a threaded nut, an expanding wedge, and a stop nut on the other. The threaded end of the rock anchor is inserted in the hole with the nut's relation to the wedge as shown in figure 4–4. After the anchoı is placed, a crowbar is inserted through the eye of the rock anchor and twisted. This causes the threads to draw the nut up against the wedge and forces the wedge out against the sides of the hole in the rock. The wedging action is strongest under a direct pull; therefore, rock anchors should always be set so that the pull is in a direct line with the shaft of the anchor. Holes for rock anchors should be drilled 5 inches deep. A 1-inch diameter drill is used for hard rock and a ¾-inch diameter drill for soft rock. The hole is drilled as neatly as possible in order that the rock anchor develops the maximum strength. In case of extremely soft rock, it is better to use some other type of anchor because the wedging action may not provide sufficient holding power.

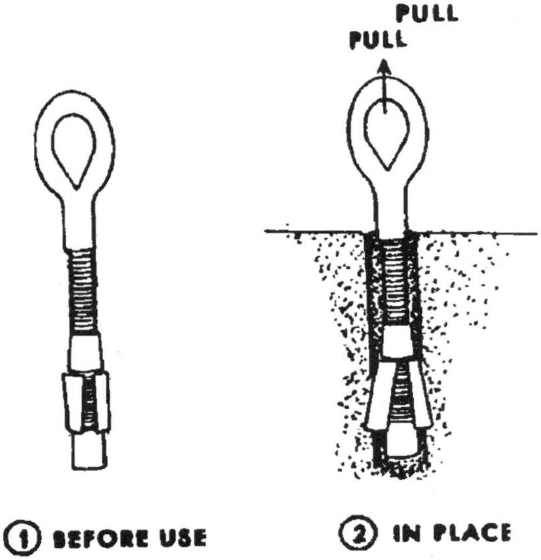

Figure 4–4. Rock anchor.

b. Picket Holdfasts.

(1) *Introduction.* A single picket, either steel or wood, can be driven into the ground as an anchor. The holding power will depend on the diameter and kind of material used, the type of soil, the depth and angle in which the picket is driven, and the angle of the guyline in relation to the ground. The holding capacities of the various types of wooden picket holdfasts are listed in table 4–1. The various picket holdfasts are shown in figure 4–5.

(2) *Single wooden picket.* Wooden stakes used for pickets should be at least 3 inches in diameter and 5 feet long. The picket is driven 3 feet into the ground at an angle of 15° from the vertical and inclined away from the direction of pull (fig. 4–6).

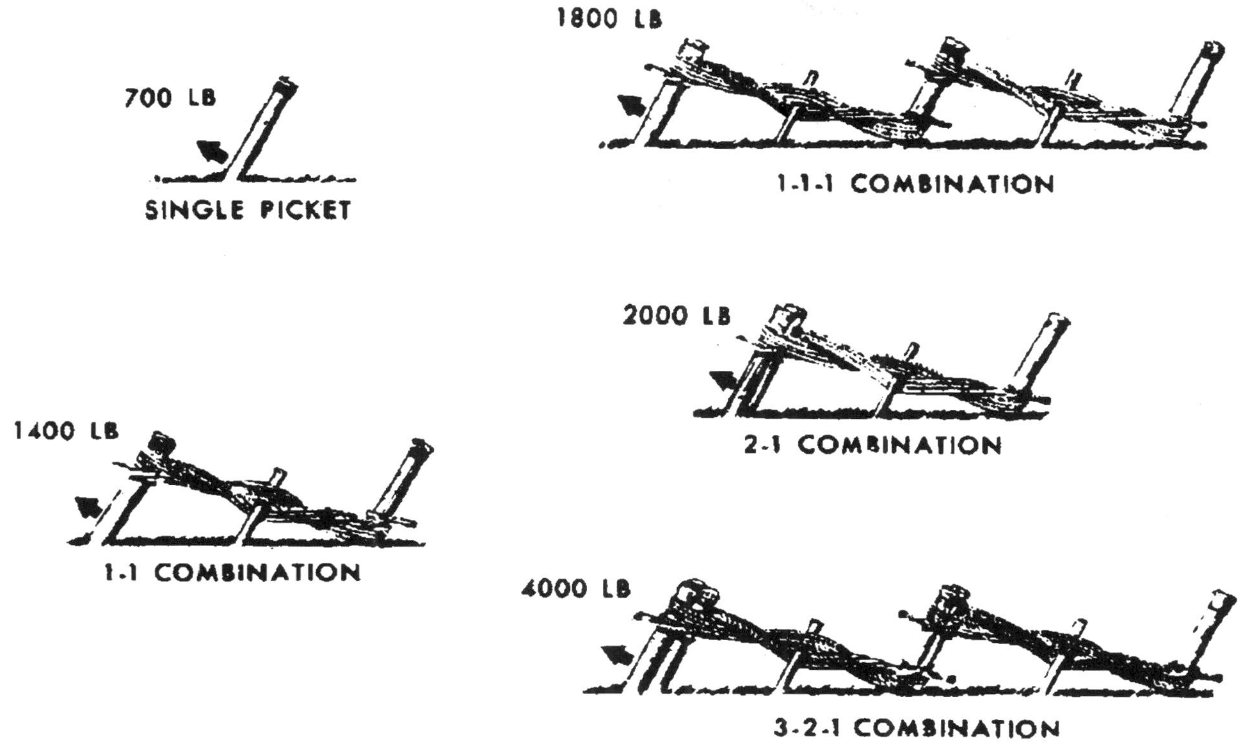

Figure 4–5. Picket holdfasts (loamy soil).

Table 4–1. Holding Power of Picket Holdfast in Loamy Soil

Holdfast	Pounds
Single picket	700
1–1 picket holdfast	1,400
1–1–1 picket holdfast	1,800
2–1 picket holdfast	2,000
3–2–1 picket holdfast	4,000

Note. Wet earth factors:
Clay and gravel mixtures 0.9
River clay and sand 0.5

(3) *Multiple wooden pickets.* The strength of a holdfast can be increased by increasing the area of the picket bearing against the ground. Two or more pickets driven into the ground, spaced 3 to 6 feet apart and lashed to-gether to distribute the load, are much stronger than a single picket. To construct the lashing, a clove hitch is tied to the top of the first picket with 4 to 6 turns around the first and second pickets, leading from the top of the first picket to the bottom of the second picket (①, fig. 4–6). Then the rope is fastened to the second picket with a clove hitch just above the turns. A stake is put between the rope turns to tighten the rope by twisting the stake and then driving it into the ground (③, fig. 4–6). This distributes the load between the pickets. If more than two pickets are used a similar lashing is made between the second and third pickets (④, fig. 4–6). If wire rope is used for lashing, only two complete turns are made around each pair of pickets. If neither fiber rope nor wire rope is available for lashing, boards may be placed from the top of the front picket to the bottom of the second picket (fig. 4–7) and nailed onto each picket. As pickets are placed farther away from the front picket, the load to the rear pickets is distributed more unevenly. Thus, the principal stength of a multiple picket holdfast is at the front pickets. The capacity of a holdfast can be increased by using two or more pickets to form the front group. This increases both the bearing surface against the soil and the breaking strength.

(4) *Steel picket holdfast.* A standard steel picket holdfast (fig. 4–8) consists of a steel box

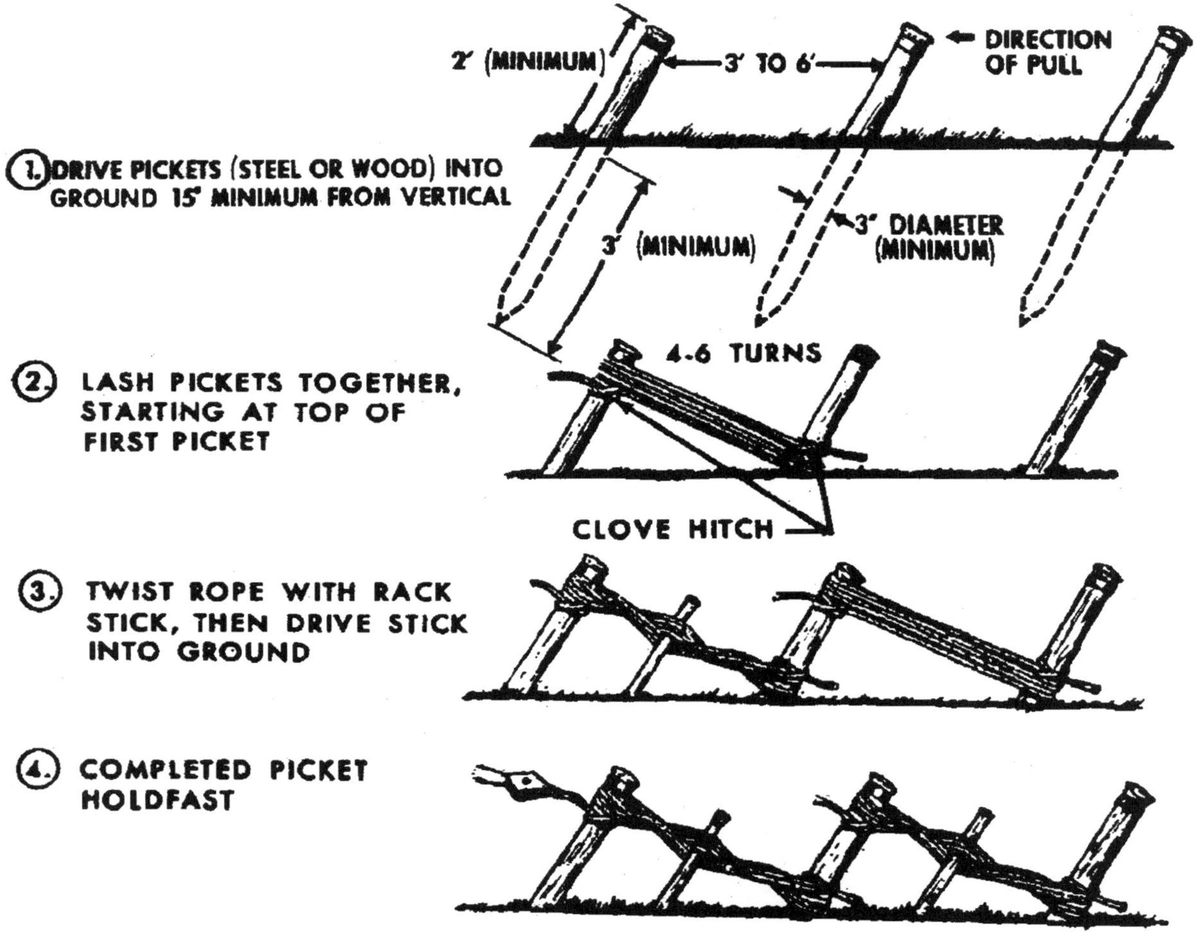

Figure 4–6. Preparing a picket holdfast.

plate with nine holes drilled through it and a steel eye welded on the end for attaching a guyline. The pickets are also steel, and are driven through the holes in a way that clinches the pickets in the ground. This holdfast is especially adapted for anchoring horizontal lines, such as the anchor cable on a ponton bridge. Two or more of these units can be used in combination to provide a stronger anchorage. A similar holdfast can be improvised with a chain by driving steel pickets through the chain links in a crisscross pattern. The rear pickets are driven in first to secure the end of the chain, and the successive pickets are installed so that there is no slack in the chain between the pickets. A lashed steel picket holdfast (fig. 4–9) consists of steel pickets lashed together with wire rope in the same way as wooden stake picket holdfasts. As an expedient, any miscellaneous light steel members can be driven into the ground and lashed together with wire rope to form an anchorage.

(5) *Rock holdfast.* A holdfast can be placed in rock by drilling into the rock and driving pickets into the holes. The pickets are lashed together with a chain (fig. 4–10). The holes are drilled about 3 feet apart, in line with the guyline. The first, or front, hole should be 2½ to 3 feet deep, and the rear hole 2 feet deep. The holes are drilled at a slight angle, inclined away from the direction of the pull.

c. Combination Holdfast. For heavy loading of an anchorage, it is desirable to spread the

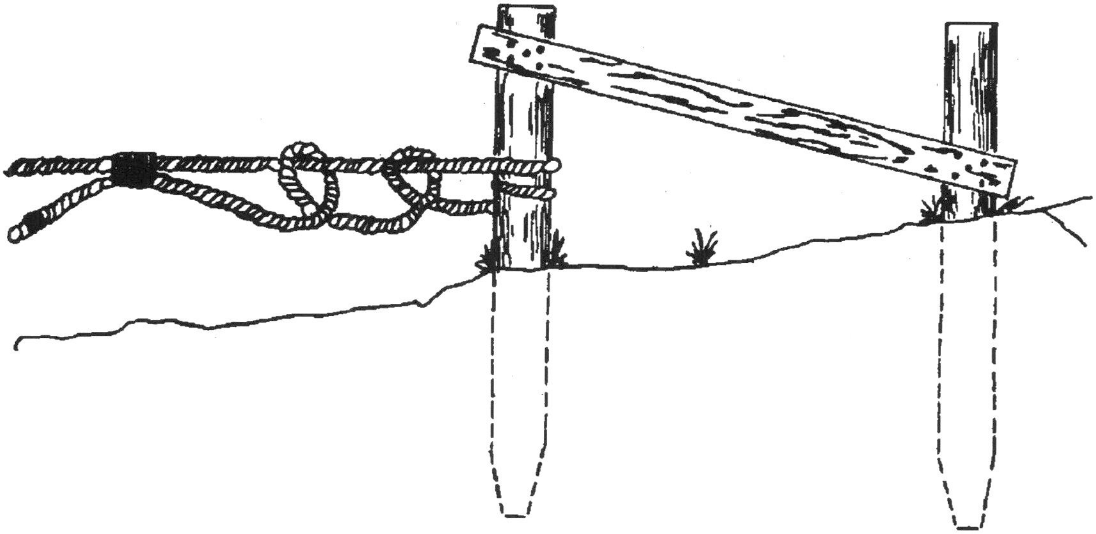

Figure 4–7. Boarded picket holdfast.

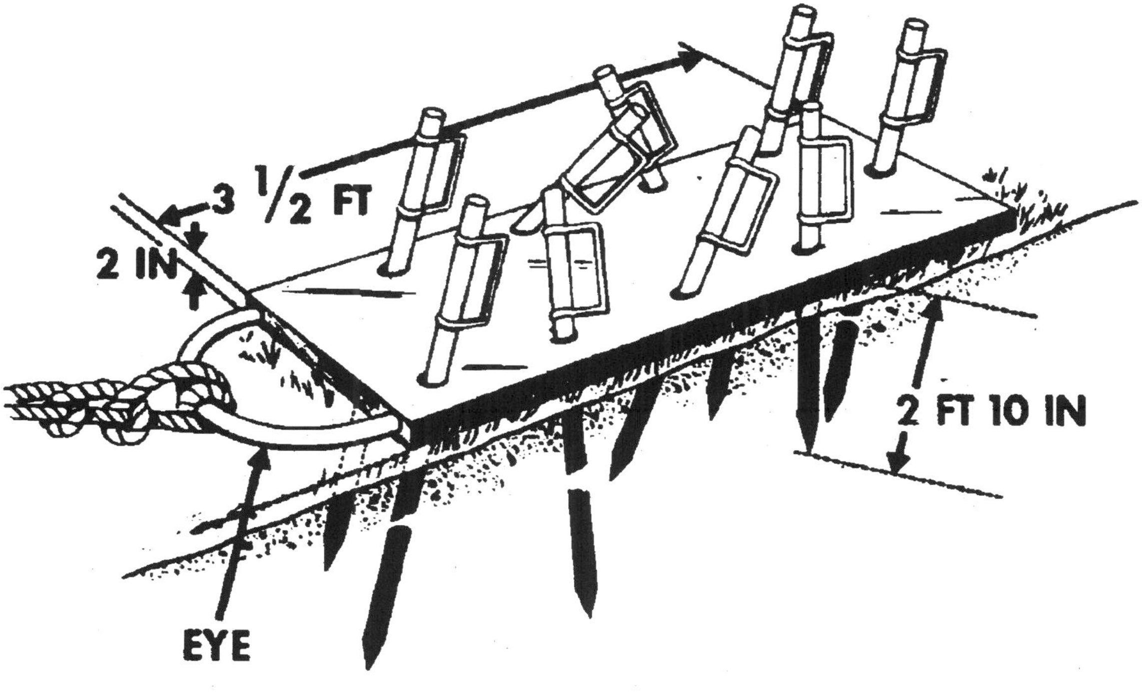

ANCHORAGE IS PROVIDED BY NINE STEEL PICKETS DRIVEN THROUGH HOLES IN THE PLATE

Figure 4–8. Standard steel-picket holdfast.

Figure 4-9. Lashed steel picket holdfast.

load over the largest possible area of ground. This can be done by increasing the number of pickets used. Four or five multiple picket holdfasts can be placed parallel to each other with a heavy log resting against the front pickets to form a combination log and picket holdfast (fig. 4-11). The guyline or anchor sling is fastened to the log which bears against the pickets. The log should bear evenly against all pickets in order to obtain maximum strength. The timber should be carefully selected to withstand the maximum pull on the line without appreciable bending. A steel crossmember can also be used to form a combination steel picket holdfast (fig. 4-12).

d. Deadman.

(1) *Construction of deadman.* A deadman is one of the best types of anchorages for heavy loads or permanent installations because of its great holding power. It may be constructed of a log, rectangular timber, steel beam or similar object buried in the ground with a guyline or sling attached to its center. This guyline or sling leads to the surface of the ground along a narrow upward sloping trench. The holding power of a deadman is affected by its frontal bearing area, its mean (average) depth, the angle of pull, the deadman material and the soil condition. The holding power increases progressively as the deadman is placed deeper and as the angle of pull approaches a horizontal position as shown in table 4-2. The holding power of a deadman must be designed to withstand the breaking strength of the line attached to it. In the construction of a deadman (fig. 4-13), a hole is dug at right-angles to the guyline and undercut 15 degrees from the vertical at the front of the hole facing the load. The guyline should be as horizontal as possible, and the sloping trench should match the slope of the guyline. The main or standing part of the line leads from the bottom of the deadman. This reduces the tendency to rotate the deadman upward out of the hole. If the line cuts into the ground, a log or board can be placed under the line at the outlet of the sloping trench. When wire rope

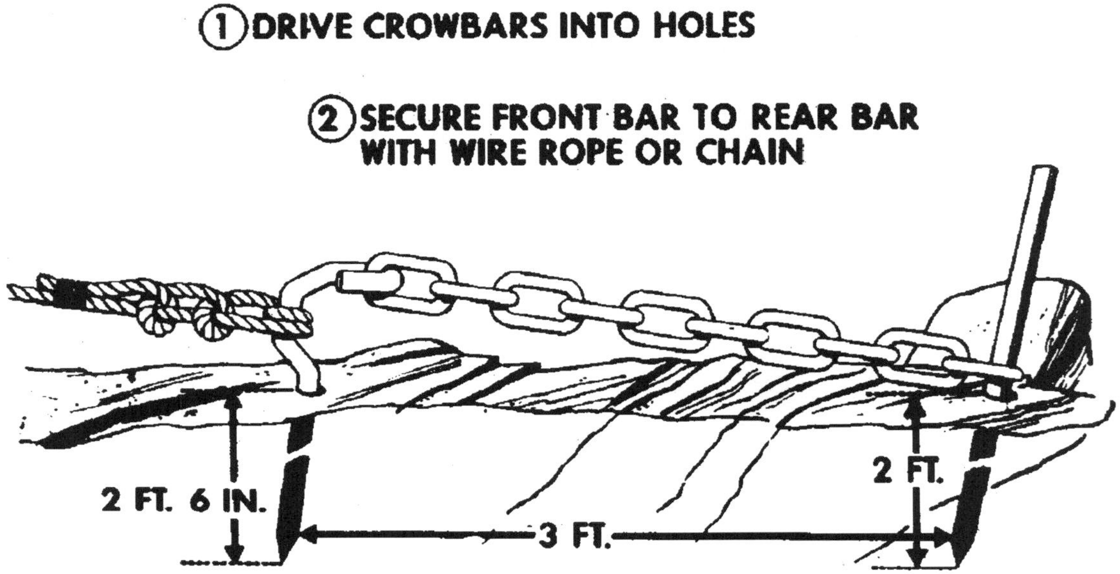

Figure 4-10. Rock holdfast.

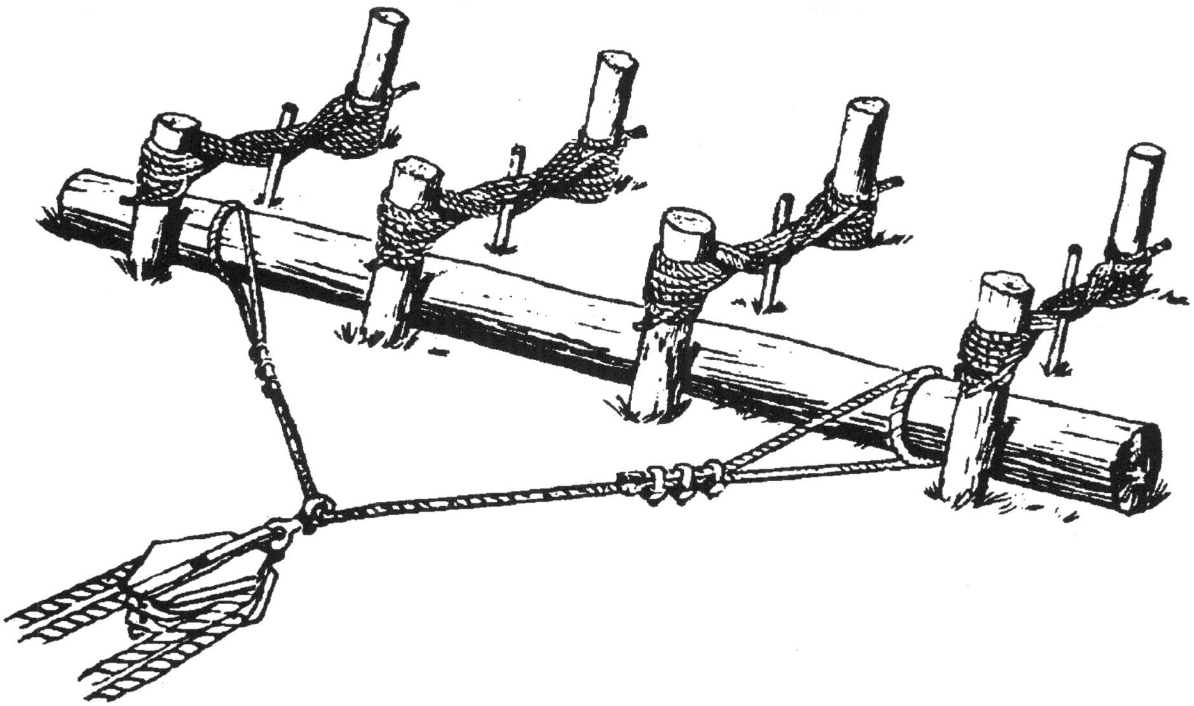

Figure 4-11. Combination log and picket holdfast.

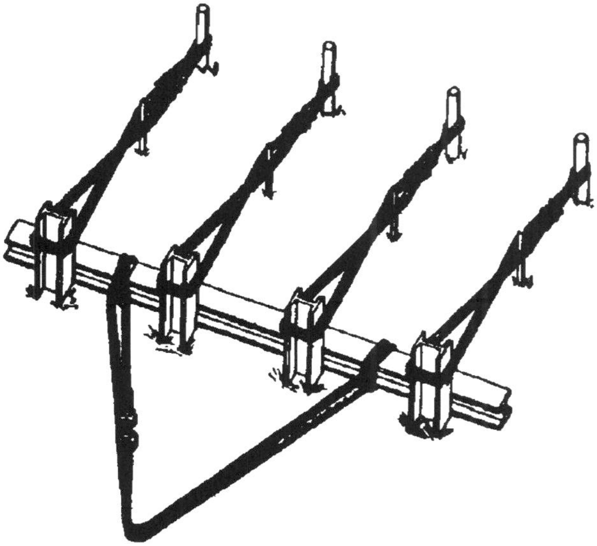

Figure 4-12. Combination steel picket holdfast.

guylines are used with a wooden deadman, a steel bearing plate should be placed on the deadman where the wire rope is attached to avoid cutting into the wood. The placement of wire rope clips should always be above the ground for retightening and maintenance.

(2) *Terms used in design of deadman.*

(*a*) *MD* = mean depth—distance from the ground level to the center of the deadman.

(*b*) *HD* = horizontal distance—distance measured horizontally from the front of the hole to the point where the sloping trench comes out of the ground.

(*c*) *VD* vertical depth—distance from ground level to the bottom of the hole.

(*d*) WST = width of sloping trench.

(*e*) *D* = timber diameter.

(*f*) *EL* = effective length—length of log that must be bearing against solid or undisturbed soil.

(*g*) *TL* = timber length—total length required.

(*h*) *HP* = holding power of deadman in ordinary earth (refer to table 4-2).

(*i*) *BS* = breaking strength of rope attached to the deadman.

(*j*) *SR* = slope ratio of guyline and sloping trench.

(*k*) BA_r = bearing area of the deadman required to hold the breaking strength of the attached rope.

Table 4-2. Holding Power of Deadmen in Ordinary Earth

Mean depth of anchorage in feet	Inclination of pull (vertical to horizontal) and safe resistance in pounds per square foot of projected area of deadmen				
	Vertical	1:1(45°)	1:2(26.5°)	1:3(18.5°)	1:4(14°)
3	600	950	1,300	1,450	1,500
4	1,050	1,750	2,200	2,600	2,700
5	1,700	2,800	3,600	4,000	4,100
6	2,400	3,800	5,100	5,800	6,000
7	3,200	5,100	7,000	8,000	8,400

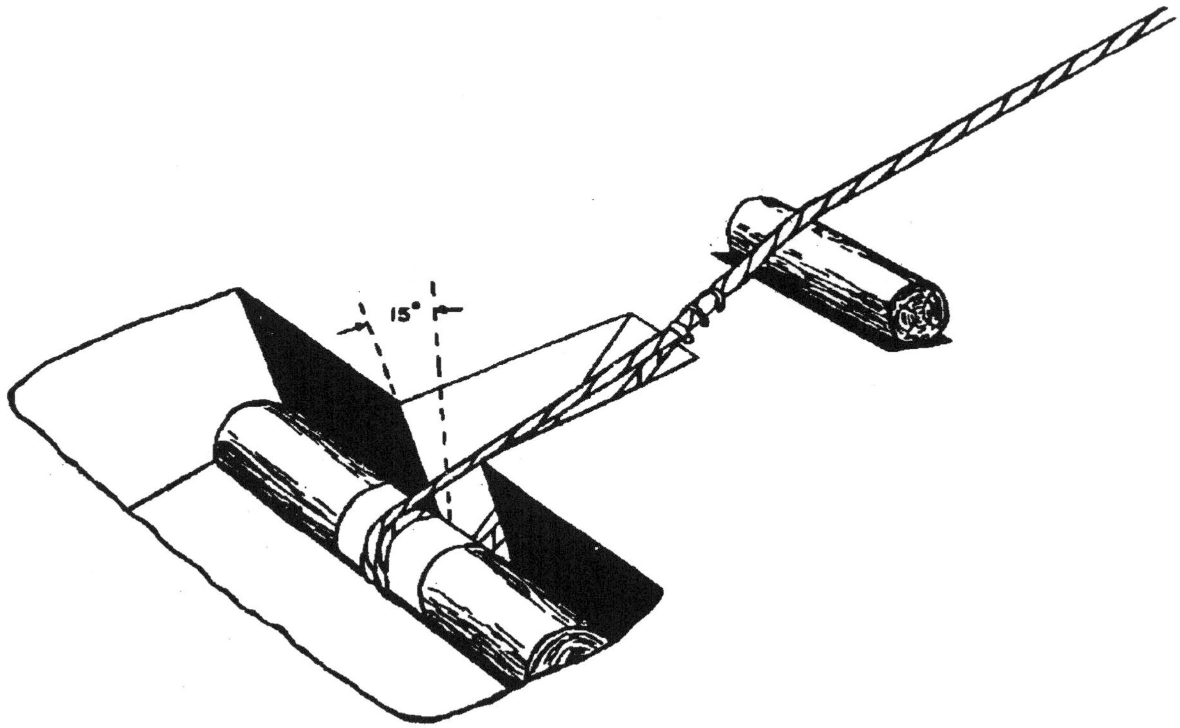

Figure 4-13. Log deadman.

(3) *Deadman formulas.*

(*a*) $BA_r = \frac{BS}{HP}$

(*b*) $EL = \frac{BA_r}{D}$

(*c*) $TL = EL + WST$

(*d*) $VD = MD + \frac{D}{2}$

(*e*) $HD = \frac{VD}{SR}$

(4) *Sample problem.*

Given:

(*a*) 1 in. dia. 6 x 19 improved plow steel rope

(*b*) Mean depth (*MD*) = 7 ft

(*c*) Slope ration (*SR*) = 1.3

(*d*) Width of sloping trench (*WST*) = 2 ft

REQUIREMENT I: Determine the length and thickness of a rectangular timber deadman if the height of face available is 18 inches (1½ ft).

(*a*) Breaking strength of wire rope (BS) = 83,600 lb. from table 1–2, chapter 1).

(*b*) Holding power of deadman (HP) = 8,000 lb/ft² (from table 4–2).

Note. The deadman should be designed to withstand a tension equal to the breaking strength of the wire rope.

(*c*) Bearing area of deadman (BA_r)

$$= \frac{BS}{HP} = \frac{83{,}600 \text{ lb}}{8{,}000 \text{ lb/ft}^2} = 10.5 \text{ ft}^2$$

(*d*) Effective length of deadman (EL)

$$= \frac{BA_r}{\text{face height}} = \frac{10.5 \text{ ft}^2}{1.5 \text{ ft}} = 7 \text{ ft}$$

(*e*) Length of deadman (TL) = EL + WST = 7 ft + 2 ft = 9 ft

(*f*) A final check to insure the rectangular timber will not fail by bending is accomplished through a length-to-thickness ratio ($L/_t$) which should be equal to or less than 9. The minimum thickness can be determined by $L/_t$ = 9 and solve for (t) : $\frac{L}{1t}$ = 9, $\frac{9}{t}$ = 9, $t = \frac{9}{9}$ = 1 ft

Thus, and 18″ x 12″ x 9′ timber is suitable.

REQUIREMENT II: Determine the length of a log deadman with a diameter of 2½ ft.

(*a*) Effective length of deadman (EL)

$$= \frac{\text{Bearing Area required } (BA_r)}{\text{Diameter of log } (D)}$$

$$\frac{10.5 \text{ ft}^2}{2.5 \text{ ft}} = 4\ 2 \text{ ft}$$

(*b*) Length of deadman (TL) = EL + WST = 4.2 ft + 2 ft = 6.2 ft

(*c*) A final check to insure that the log will not fail in bending is accomplished through a length-to-diameter ratio ($L/_d$) which should be equal to or less than 5. The ratio for requirement II would be equal to $L/_d$ = 6.2/2.5 = 2.5. This is less than 5, therefore the log will not fail by bending.

(5) *Length-to-diameter ratio.* If the length-to-diameter ratios for a log or a rectangular timber are exceeded, this means the length requirements must be decreased. This can be accomplished by one of the following methods:

(*a*) Increase the mean depth.

(*b*) Increase the slope ration (the guyline becomes more horizontal).

(*c*) Increase the thickness of the deadman.

(*d*) Decrease the width of the sloping tranch, if possible.

4–4. Using the Nomograph to Design Deadmen

Nomographs and charts have been prepared to facilitate the design of deadmen in the field. The deadmen are designed to resist the breaking strength of the cable. The required length and thickness are based on allowable soil bearing with 1 foot of length added to compensate for the width of the cable trench. The required thickness is based on an ($L/_d$) ratio of *8* for logs and an ($L/_t$) ration of 9 for cut timber.

a. Log Deadmen.

Sample problem.

Given: ¾-inch IPS cable. Required deadman to be buried 5 feet at a slope of ¼.

Solution: With this information use the nomograph (fig. 4–14) to determine the diameter and length of the deadman required. Figure 4–15 shows the steps graphically on an incomplete nomograph. Lay a straightedge across section A–A (left-hand scale) on the 5 foot depth at deadman and ¼ slope, and on ¾-inch IPS on B–B. Read across the straightedge and locate a point on section C–C. Then go horizontally across the graph and intersect the diameter of the log deadmen available. Assume a 30-inch diameter log is available. Go vertically up from the intersection on the log and read the length of deadman required. In this case the deadman must be over 5½ feet long. Care must be taken not to select a log deadman in the darkened area of the nomograph because in this area the log will fail by bending.

b. Rectangular Timber Deadman.

Sample problem.

Given: ¾-inch IPS cable. Deadman to be buried 5 feet at a slope of ¼.

Solution: Use the same ¼ slope and 5 foot depth, along with the procedure to the left of the graph (fig. 4–14) as used in *a* above. At C–C go horizontally across the graph to the timber with an 18-inch face. Reading down (working with cut timber) it can be seen that the length is 8 feet 6 inches, and the minimum thickness is 11.5 inches. None of the timber

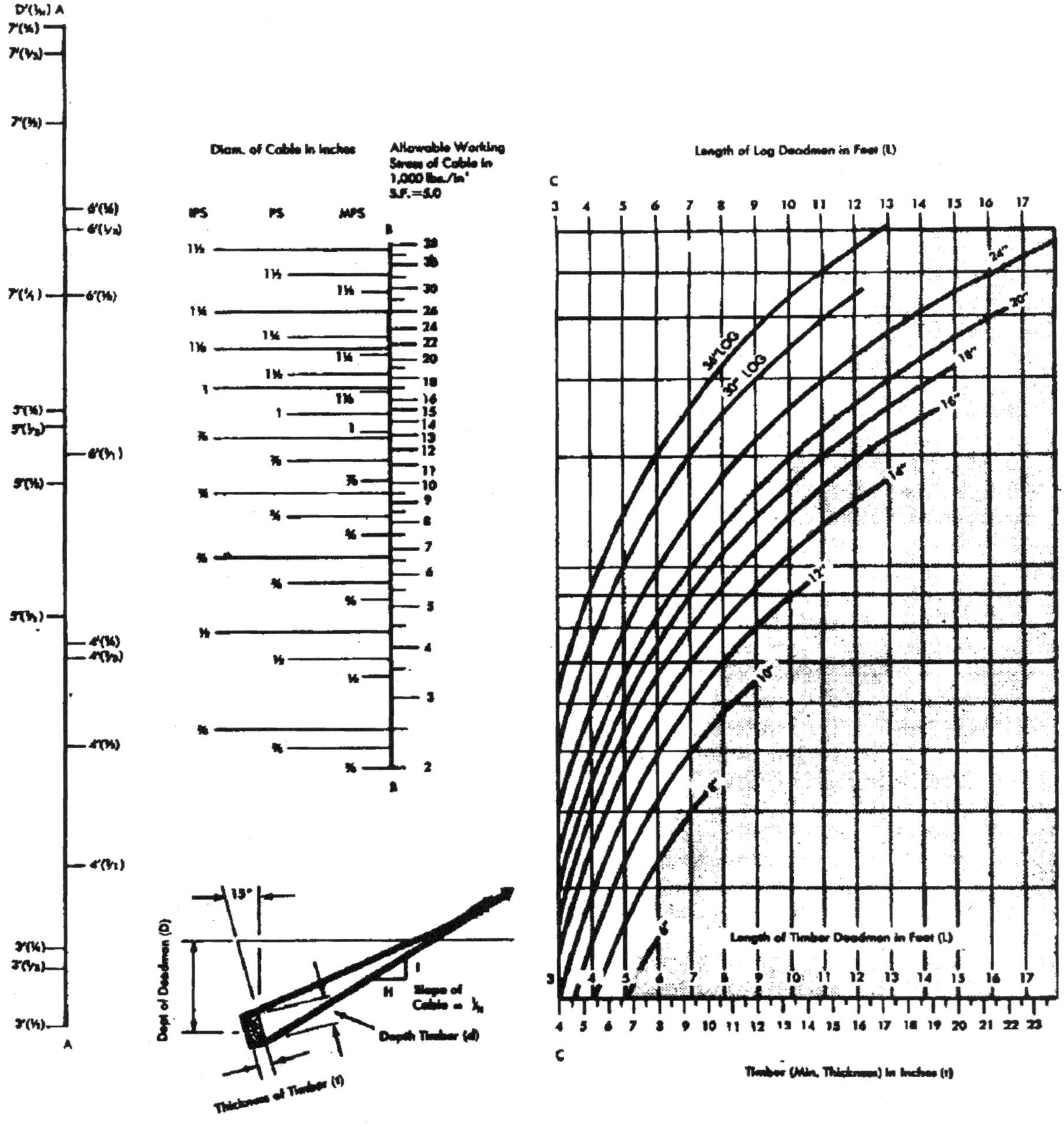

Figure 4–14. Design of deadmen.

sizes shown on the nomograph will fail due to bending.

c. Horizontal Distance. The distance behind the tower in which deadmen are placed is determined by the following formula:

$$\text{Horizontal distance} = \frac{\text{tower height} + \text{deadman depth}}{\text{slope ratio}}$$

Sample problem:

Given: The tower height is 25 feet 4¼ inches, and the deadman depth is 7 feet with a 1 : 4 slope.

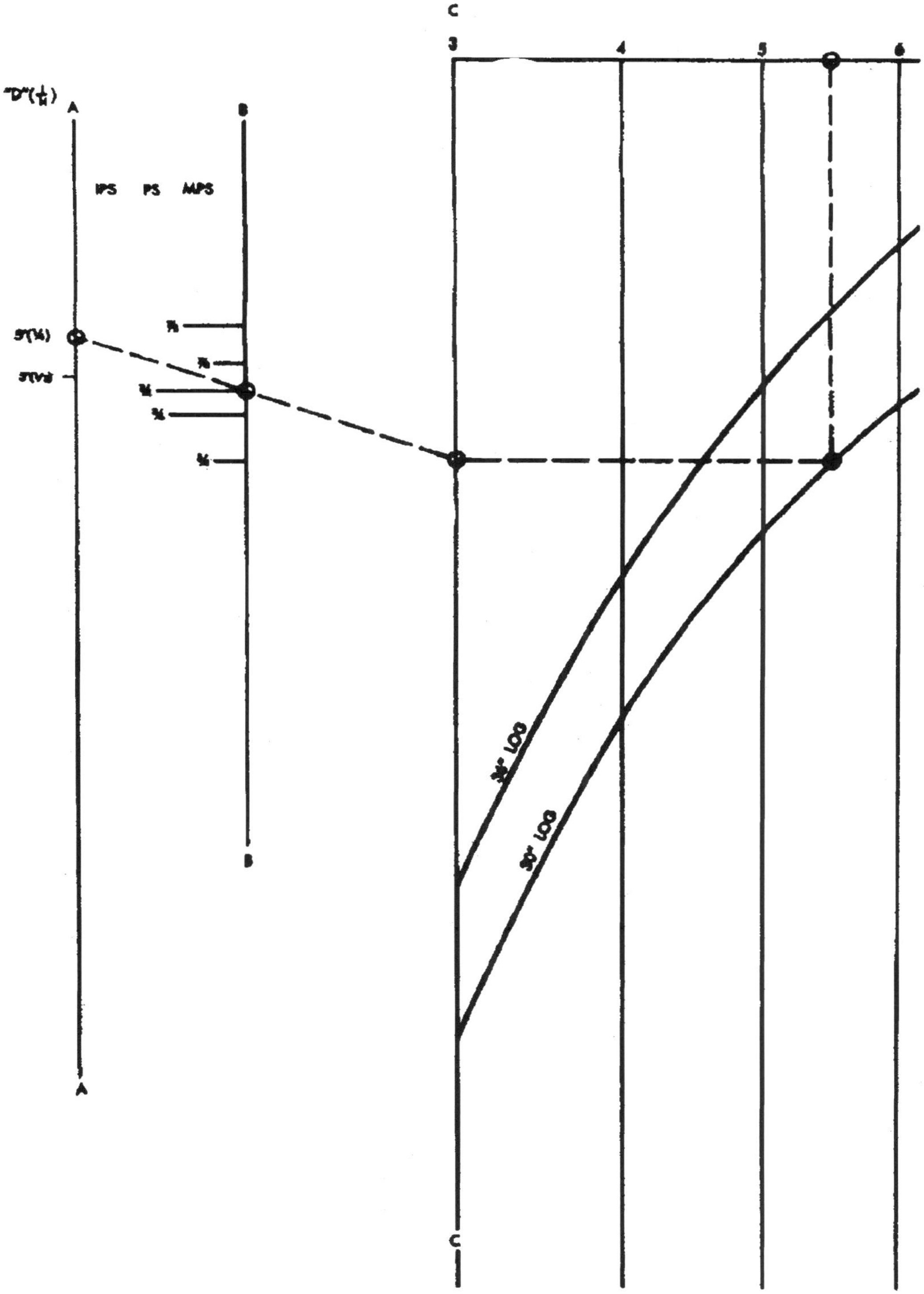

Figure 4–15. Use of nomograph.

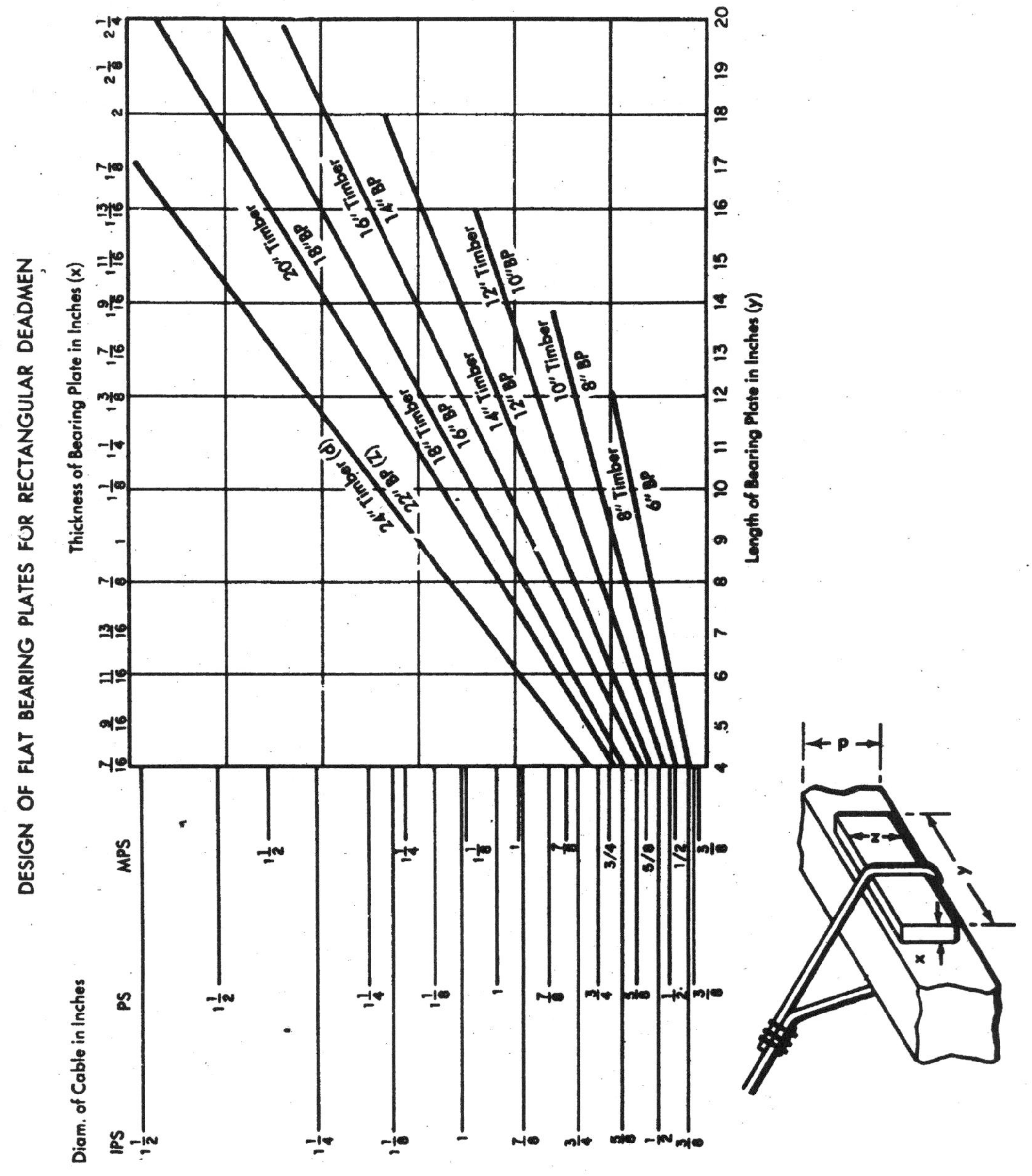

Figure 4–16. Design of flat bearing plate.

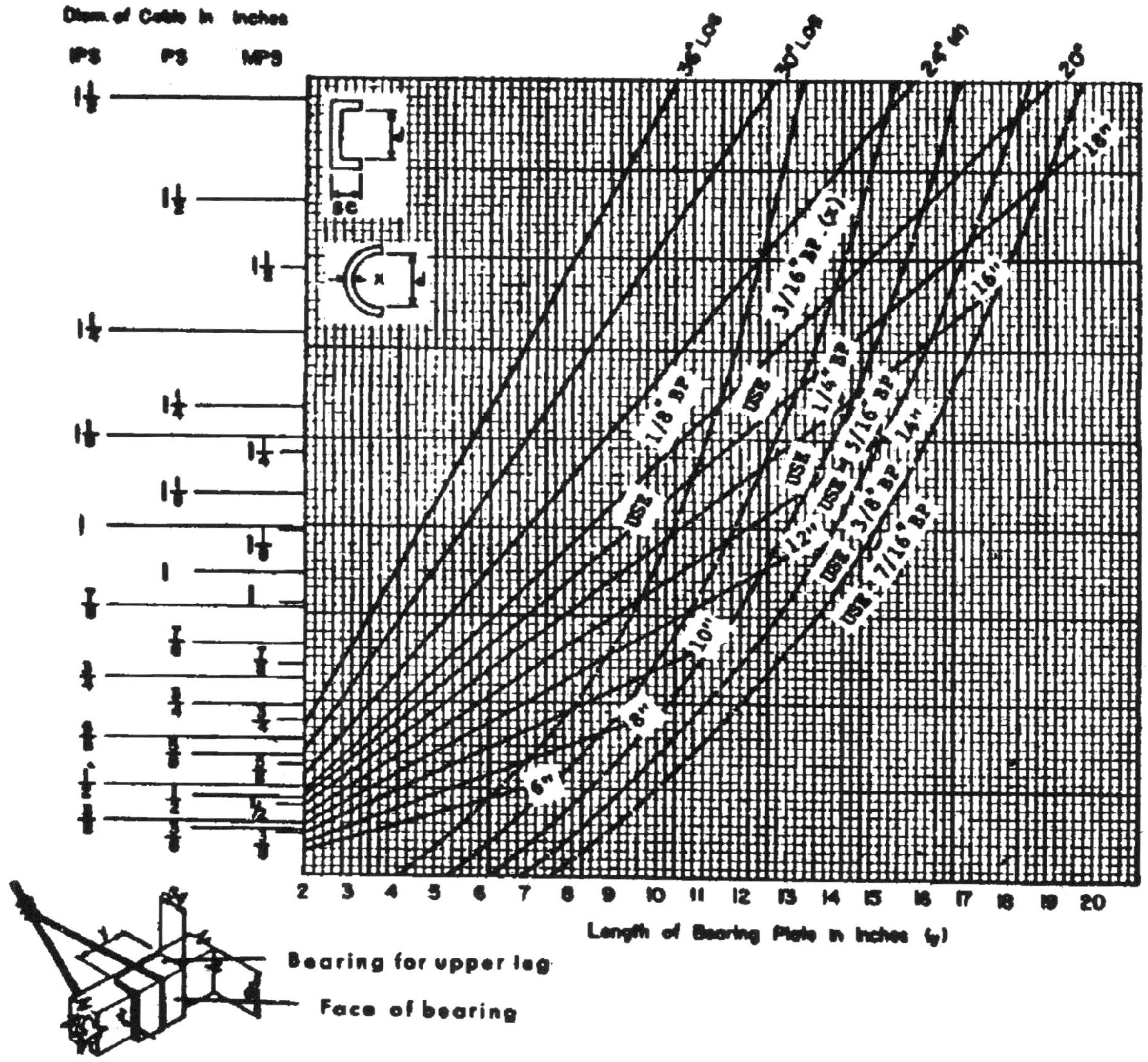

Figure 4 17. Design of formed bearing plate.

$$\textit{Solution:}\quad \frac{25\text{ ft }4\frac{1}{4} + 7\text{ ft}}{1/4}$$

$$= \frac{32\text{ ft }4\frac{1}{4}\text{ in}}{1/4} = 129\text{ ft }5\text{ in.}$$

Place deadman 129 ft behind tower.

Note. Horizontal distance without tower

$$= \frac{\text{deadman depth}}{\text{slope ration}} = \frac{7\text{ ft}}{1/4} = 28\text{ ft.}$$

4–5. Bearing Plates

To prevent the cable from cutting into the wood, place a metal bearing plate on the deadman. There are two types of bearing plates, the flat bearing plate and the formed bearing plate, each with its particular advantages. The flat is easily fabricated, while the shaped or formed can be made of much thinner steel. A sample problem in the design of bearing plates is given below.

a. Design of Flat Bearing Plate.

Sample problem.

Given: 12 in. x 12 in. timber

¾ in IPS cable

Solution: Enter the graph (fig. 4–16) from left of ¾-inch cable and go horizontally across graph to intersect line marked 12" timber, which shows that the plate will be 10 inches wide. (The bearing plate is made 2 inches narrower than the timber to prevent cutting into the anchor cable.) Drop vertically and determine the length of the plate, which is 9½ inches. Go to the top vertically along the line to where it intersects with ¾" cable and determine the minimum required thickness, 1 1/16 inches. Thus the necessary bearing plate must be 1 1/16 inches x 9½ inches x 10 inches.

b. Design of Formed Bearing Plate. The formed bearing plates are either curved to fit logs or formed to fit rectangular timber. In the case of a log, the bearing plate must go one-half way (180°) around the log. For a shaped timber, the bearing plate extends the depth of the timber with an extended portion at the top and the bottom (fig. 4–17). Each extended portion should be one-half the depth of the timber.

Sample problem.

Given: 14-in. log or timber with 14-in. face. 1 1/8 MPS cable.

Solution: Design a formed bearing plate. Enter graph (fig. 4–17) on left at 1⅛ MPS and go horizontally across to intersect the 14" line. Note that lines intersect in an area requiring a ¼-inch plate. Drop vertically to the bottom of the graph to determine the length of the plate, which in this instance is 12 inches. If a log is used, the width of the bearing plate is equal to ½ the circumference of the log, $\frac{\pi d}{2}$ or in this case 22 inches. $\frac{\pi d}{2} = \frac{3.14 \times 14}{2} = 21.98$, use 22 inches. The bearing plate would therefore be ¼ inch x 12 inches x 22 inches. For a rectangular timber, the width of the plate would be 14 inches for the face and 7 inches for the width of each leg, or a total width of 28 inches (sketch on fig. 4–17). The bearing plate would therefore be ¼ inch x 12 inches x 28 inches.

Section II. GUYLINES

4–6. Introduction

Guylines are ropes or chains attached to an object to steady, guide, or secure it. The lines leading from the object or structure are attached to an anchor system (fig. 4–18). When a load is applied to the structure supported by the guylines, a portion of the load is passed through each supporting guyline to its anchor. The amount of tension on a guyline will depend on the main load, the position and weight of the structure, the alinement of the guyline with the structure and the main load, and the angle of the guyline. For example, if the supported structure is vertical, the stress on each guyline is very small, but if the angle of the structure is 45°, the stress on the guylines supporting the structure will increase considerably. Wire rope is preferred for guylines because of its strength and resistance to corrosion. Fiber is also used for guylines, particularly on temporary structures. The number and size of guylines required depends on the type of structure to be supported and the tension or pull exerted on the guylines while the structure is being used.

4–7. Number of Guylines

Usually a minimum of four guylines are used for ginpoles and boom derricks and two for shears. The guylines should be evenly spaced around the structure. In a long slender structure it is sometimes necessary to provide support at several points in a tiered effect. In such cases, there might be four guylines from the center of a long pole to anchorage on the ground and four additional guylines from the top of the pole to anchorage on the ground.

4–8. Tension

The tension that will be exerted on the guylines must be determined beforehand in order to select the proper size and material to be used. The maximum load or tension on a guyline will result when a guyline is in direct line with the load and the structure. This tension should be considered in all strength calculations of guylines. The following is the formula for determining this tension for ginpoles and shears (fig. 4–19):

$$T = \frac{(W_1 + \frac{1}{2} W_2)D}{Y}$$

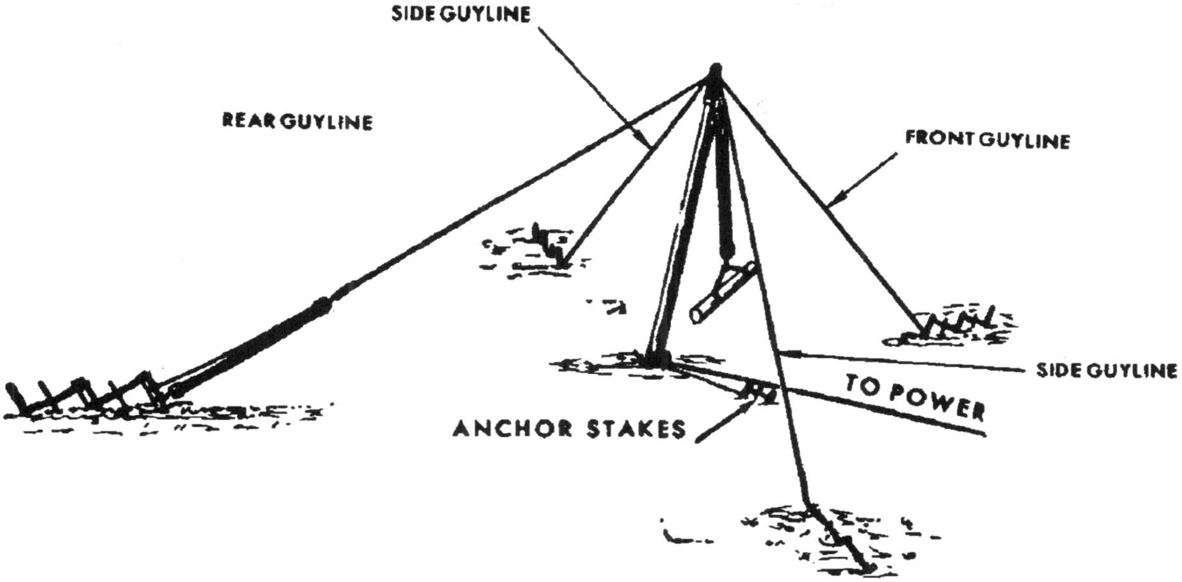

Figure 4–18. Typical guyline installations.

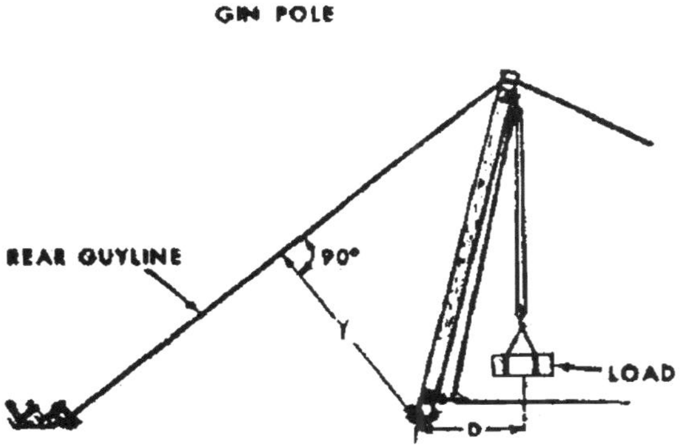

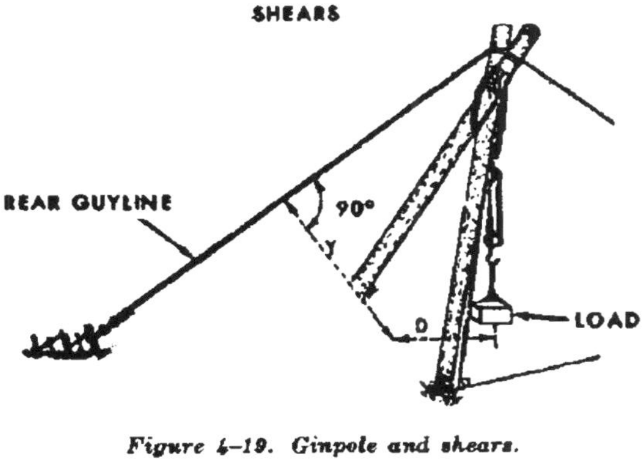

Figure 4–19. Ginpole and shears.

T = Tension in guyline
W_L = Weight of the load
W_s = Weight of spar or spars
D = Drift distance, measured from the base of the ginpole or shears to the center of the suspended load along the ground.
Y = Perpendicular distance from the rear guyline to the base of the gin pole or for a shears, to a point on the ground midway between the shearlegs.

Sample problem.

REQUIREMENT I: *Gin Pole:*

a. *Given:* Load (W_L) = 2,400 lb
Weight of spar (W_s) = 800 lb
Drift distance (D) = 20 ft

b. *Solution:* $T = \frac{(W_L + \frac{1}{2}W_s)\,D}{Y}$

$$\frac{(2400 + \frac{1}{2}\,(800))\;20}{28} = 2{,}000 \text{ lb tension}$$

in the rear or supporting guyline.

REQUIREMENT II: *Shears:*

a. *Given:* The same conditions exist as in requirement I, except that there are *two* spars, each one weighing 800 lb.

b. Solution:

$$T = \frac{(W_L + \frac{1}{2} W_s) D}{Y} = \frac{(2400 + \frac{1}{2}(800 + 800))\ 20}{Y}$$

$$= 2{,}285 \text{ lb}$$

Note. The reason the shears produced a greater tension in the rear guyline was due to the weight of an additional spar.

4–9. Size of Guyline

The size of the guyline to be used will depend on the amount of tension to be placed on it. Since the tension on a guyline may be affected by shock loading, and its strength affected by knots, sharp bends, age, and condition, the appropriate safety factors must be incorporated. Therefore, a rope chosen for the guyline should have a safe working capacity equal to or greater than the tension placed on the guyline.

4–10. Anchorage Requirements

An ideal anchorage system should be designed to withstand a tension equal to the breaking strength of the guyline attached to it. If a 3/8-inch diameter manila rope is used as a guyline, the anchorage used must have the capability of withstanding a tension of 1,350 pounds which is the breaking strength of the 3/8-inch diameter manila rope. If picket holdfasts are used, it would require at least a 1–1 combination (1,400 lb capacity in ordinary soil). The guyline should be anchored as far as possible from the base of the installation to obtain a greater holding power from the anchorage system. The recommended minimum distance from the base of the installation to the anchorage for the guyline is twice the height of the installation.

CHAPTER 5

LIFTING AND MOVING LOADS

Section I. LIFTING EQUIPMENT

5–1. Gin Pole

A gin pole consists of an upright spar which is guyed at the top to maintain it in a vertical or nearly vertical position and is equipped with suitable hoisting tackle. The vertical spar may be of timber, a wide-flange steel beam section, a railroad rail, or similar members of sufficient strength to support the load being lifted. The load may be hoisted by hand tackle or by use of hand- or engine-driven hoists. The gin pole is used widely in erection work because of the ease with which it can be rigged, moved, and operated. It is suitable for raising loads of medium weight to heights of 10 to 50 feet where only a vertical lift is required. The gin pole may also be used to drag loads-horizontally toward the base of the pole in preparation for a vertical lift. It cannot be drifted (inclined) more than 45 degrees from the vertical or 7/10 the height of the pole, nor is it suitable for swinging the load horizontally. The length and thickness of the gin pole depends on the purpose for which it is installed. It should not be longer than 60 times its minimum thickness because of the tendency to buckle under compression. A usable rule is to allow five feet of pole for each inch of minimum thickness. Table 5–1 lists values for the use of spruce timbers as gin poles, with allowance for normal stresses in hoisting operations.

Table 5–1. Safe Capacity of Spruce Timber as Gin Poles in Normal Operations

Size of timber in inches	Safe capacity in pounds for given length of timber					
	20 feet	25 feet	30 feet	40 feet	50 feet	60 feet
6 dia	5,000	3,000	2,000			
8 dia		11,000	8,000	5,000	3,000	
10 dia	31,000	24,000	16,000	9,000	6,000	
12 dia			31,000	19,000	12,000	9,000
6 x 6	6,000	4,000	3,000			
8 x 8		14,000	10,000	6,000	4,000	
10 x 10	40,000	30,000	20,000	12,000	8,000	
12 x 12			40,000	24,000	16,000	12,000

Note. Safe capacity of each le of shears or tripod is seven-eighths of the value given for a gin pole.

a. Rigging. In rigging a gin pole, lay out the pole with the base at the spot where it is to be erected. In order to make provisions for the guylines and tackle blocks, place the gin pole on cribbing for ease of lashing. Figure 4–18 shows the lashing on top of a gin pole and the method of attaching guys. The procedure is as follows:

(1) Make a tight lashing of eight turns of fiber rope about 1 foot from the top of the pole, with two of the center turns engaging the hook of the upper block of the tackle. Secure the ends of the lashing with a square knot. Nail wooden cleats (boards) to the pole flush with the lower and upper sides of the lashing to prevent the lashing from slipping.

(2) Lay out guy ropes, each four times the length of the gin pole. In the center of each guy rope, form a clove hitch over the top of the pole next to the tackle lashing (fig. 5–1), and

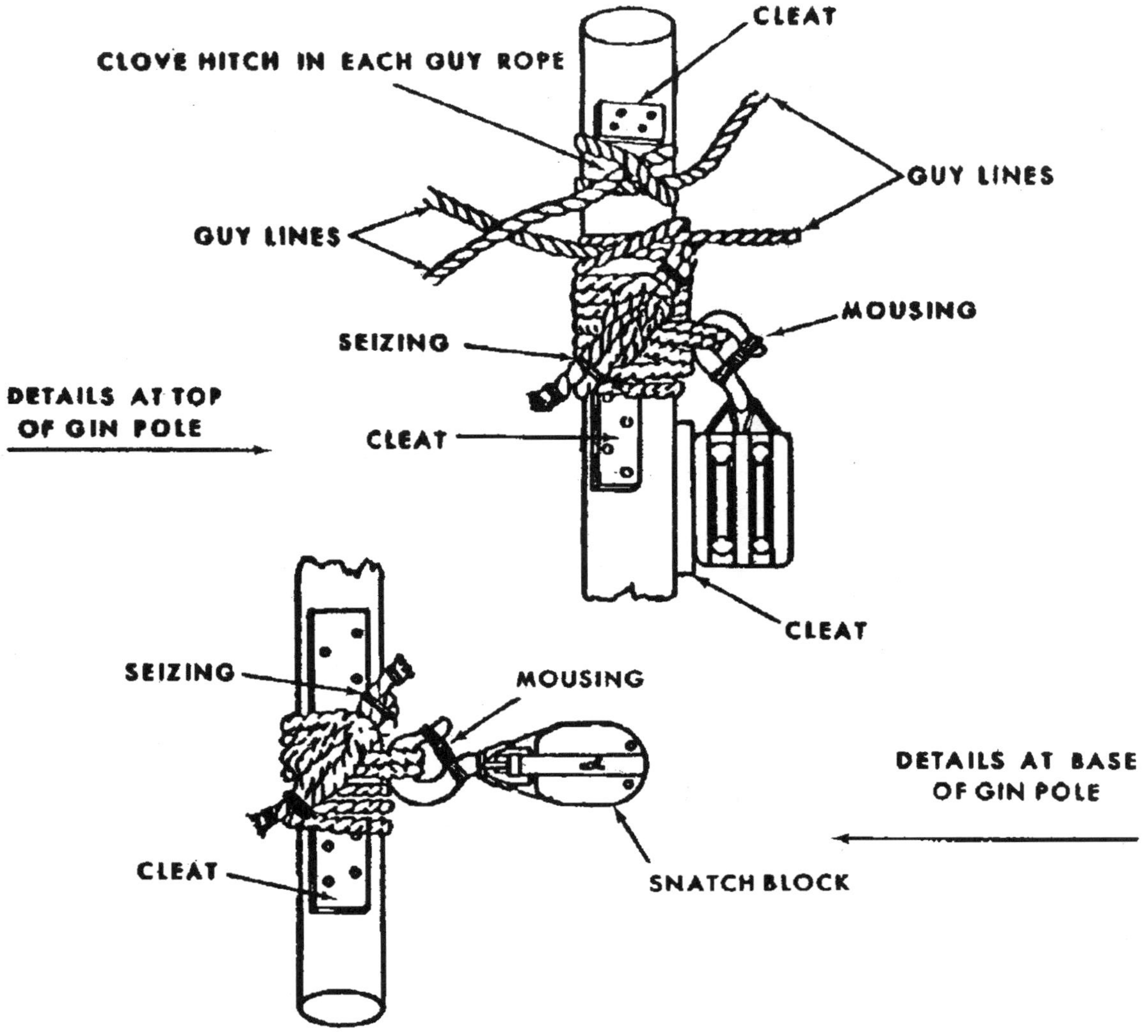

Figure 5–1. Lashing for a gin pole.

be sure the guylines are alined in the direction of their anchors.

(3) Lash a block to the gin pole about 2 feet from the base of the pole, the same as was done for the tackle lashing at the top, and place a cleat above the lashing to prevent slipping. This block serves as a leading block on the fall line which allows a directional change of pull from the vertical to the horizontal. A snatch block is the most convenient type to use for this purpose.

(4) Reeve the hoisting tackle and use the block lashed to the top of the pole so that the fall line can be passed through the leading block at the base of the gin pole.

(5) Drive a stake about 3 feet from the base of the gin pole. Tie a rope from the stake to the base of the pole below the lashing on the leading block and near the bottom of the pole. This is to prevent the pole from skidding while it is being erected.

(6) Check all lines to be sure that they are not snarled. Check all lashings to see that they are made up properly, and see that all knots are tight. Check the hooks on the blocks to see that they are moused properly. The gin pole is now ready to be erected.

b. Erecting. A gin pole 40 feet long may be raised easily by hand, but longer poles must be raised by supplementary rigging or power equipment. Figure 5–2 shows a gin pole being erected. The number of men needed depends on the weight of the pole. The procedure is as follows:

(1) Dig a hole about 2 feet deep for the base of the gin pole.

(2) String out the guys to their respective anchorages and assign a man to each anchorage to control the slack in the guyline with a round turn around the anchorage as the pole is raised. If it has not been done already, install an anchorage for the base of the pole.

(3) If necessary, the tackle system utilized to raise and lower the load may be used to assist in raising the gin pole, but the attaching of an additional tackle system to the rear guyline is preferable. Attach the running block of the rear guyline tackle system (fig 4–18) to the rear guyline the end of which is at this point of erection near the base of the gin pole. The fixed or stationary block is then secured to the rear anchor. The fall line should come out of the running block to give greater mechanical advantage to the tackle system. The tackle system is stretched to the base of the pole before it is erected to prevent the chocking of the tackle blocks during the erection of the gin pole.

(4) Keep a slight tension on the rear guyline and *on each of the side guylines,* haul in on the fall line of the tackle system, while eight men (more for larger poles) raise the top of the pole by hand (fig. 5–2) until the tackle system can take control.

(5) The rear guyline must be kept under tension to prevent the pole from swinging and throwing all of its weight on one of the side guys.

(6) When the pole is in its final position, approximately vertical or inclined as desired, make all guys fast to their anchorages with the round turn and two half hitches. It frequently is desirable to double the portion of rope used for the half hitches.

(7) Open the leading block at the base of the gin pole and place the fall line from the tackle system through it. When the leading block is closed the gin pole is ready for use. If it is necessary to move (drift) the top of the pole without moving the base, it should be done when there is no load on the pole, unless the guys are equipped with tackle.

c. Operating. The gin pole is particularly adapted to vertical lifts (fig. 5–3). It also is

Figure 5–2. Erecting a gin pole.

Figure 5–3. Hoisting with a gin pole.

used under some circumstances for lifting and pulling at the same time so that the load being moved travels toward the gin pole just off the ground. When used in this manner, a snubbing line of some kind must be attached to the other end of the load being dragged and kept under tension at all times. Tag lines should be used to control loads being lifted vertically. A tag line is a light line fastened to one end of the load and kept under slight tension during hoisting.

5–2. Tripod

A tripod consists of three legs lashed or secured at the top. The advantage of the tripod over other rigging installations is its stability and that it requires no guylines to hold it in place. Its disadvantage is that the load can be moved only up and down. The load capacity of a tripod is approximately one and one-half times that of shears made of the same size material.

a. Rigging. There are two methods of lashing a tripod, either of which is suitable provided the lashing material is strong enough. The material used for lashing can be fiber rope, wire rope, or chain. Metal rings joined with short chain sections and large enough to slip over the top of the tripod legs also can be used. The method described in par (1) below is for fiber rope 1 inch in diameter or smaller. Since the strength of the tripod is affected directly by the strength of the rope and the lashing used, more turns than described below should be used for extra heavy loads and fewer turns can be used for light loads.

(1) *Procedure.*

(*a*) Select three spars of approximately equal size and place a mark near the top of each spar to indicate the center of the lashing.

(*b*) Lay two of the spars parallel with their tops resting on a skid or block and a third spar between the first two, with the butt in the opposite direction and the lashing marks on all three in line. The spacing between spars should be about one-half the diameter of the spars. Leave the space between the spars so that the lashing will not be drawn too tight when the tripod is erected.

(*c*) With a 1-inch rope, make a clove hitch around one of the outside spars about 4 inches above the lashing mark and take eight turns of the line are

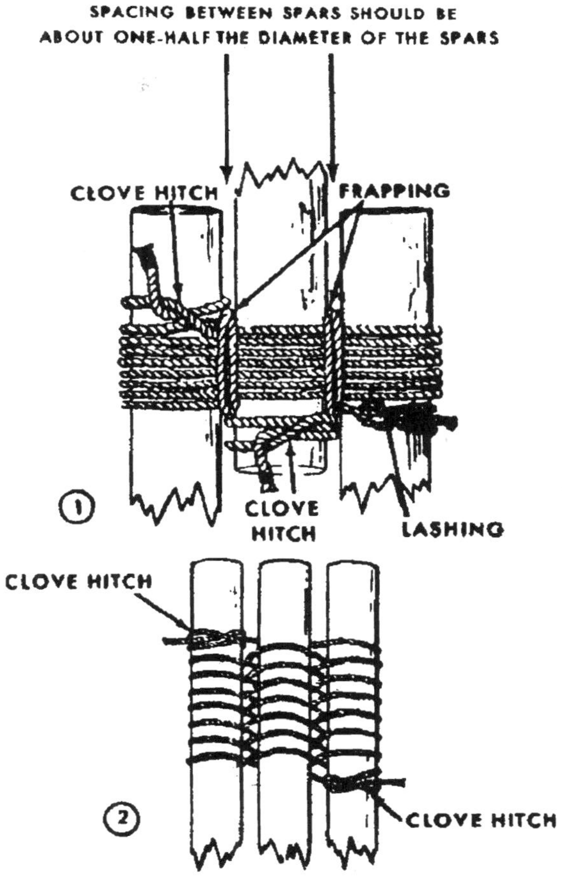

Figure 5–4. Lashing for a tripod.

fig. 5–4). Be sure to maintain the space between the spars while making the turns.

(*d*) Finish the lashing by taking two close frapping turns around the lashing between each pair of spars. Secure the end of the rope with a clove hitch on the center spar just above the lashing. Frapping turns should not be drawn too tight.

(2) *Alternate procedure.*

(*a*) An alternate procedure (②, fig. 5–4) can be used when slender poles not more than 20 feet long are being used, or when some means other than hand power is available for erection.

(*b*) Lay the three spars parallel to each other with an interval between them slightly greater than twice the diameter of the rope to be used. Rest the tops of the poles on a skid so that the ends project over the skid approximately 2 feet and the butts of the three spars are in line.

(*c*) Put a clove hitch on one outside leg at the bottom of the position the lashing will occupy which is approximately 2 feet from the end. Weave the line over the middle leg, under and around the outer leg, under the middle leg, over and around the first leg, and continue this weaving for eight turns. Finish with a clove hitch on the outer leg.

b. Erecting. The legs of a tripod in its final position should be spread so that each leg is equidistant (fig. 5-5) from the others. This spread should not be less than one-half nor more than two-thirds of the length of the legs. Chain, rope, or boards should be used to hold the legs in this position. A leading block for the fall line of the tackle may be lashed to one of the legs. The procedure is as follows:

(1) Raise the tops of the spars about 4 feet, keeping the base of the legs on the ground.

(2) Cross the two outer legs. The third or center leg then rests on top of the cross. With the legs in this position, pass a sling over the cross so that it passes over the top or center leg and around the other two.

(3) Hook the upper block of a tackle to the sling and mouse the hook.

(4) Continue raising the tripod by pushing in on the legs as they are lifted at the center. Eight men should be able to raise an ordinary tripod into position.

(5) When the tripod legs are in their final position, place a rope or chain lashing between the legs to hold them from shifting.

c. Erecting Large Tripods. For larger tripod installations it may be necessary to erect a small gin pole to raise the tripod into position. Tripods lashed in the manner described in *a* above with the three legs laid together, must be erected by raising the tops of the legs until the legs clear the gound so they can be spread apart. Guylines or tag lines should be used to assist in steadying the legs while they are being raised. The outer legs should be crossed so that the center leg is on top of the cross, and the sling for the hoisting tackle should pass over the center leg and around the two outer legs at the cross.

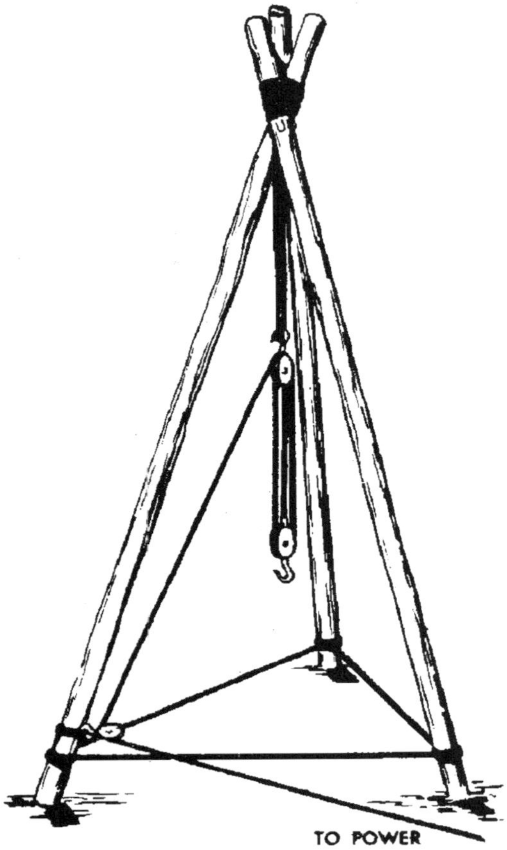

Figure 5-5. Tripod assembled for use.

5-3. Shears

Shears made by lashing two legs together with a rope is well adapted for lifting heavy machinery or other bulky loads. It is formed by two members crossed at their tops, with the hoisting tackle suspended from the intersection. The shears must be guyed to hold it in position. The shears is quickly assembled and erected. It requires only two guys, and is adapted to working at an inclination from the vertical. The shear legs may be round poles, timbers, heavy planks, or steel bars, depending on the material at hand and the purpose of the shears. In determining the size of the members to be used, the load to be lifted and the ratio of the length and diameter of the legs are the determining factors. For heavy loads the length-diameter (L/d) ratio should not exceed

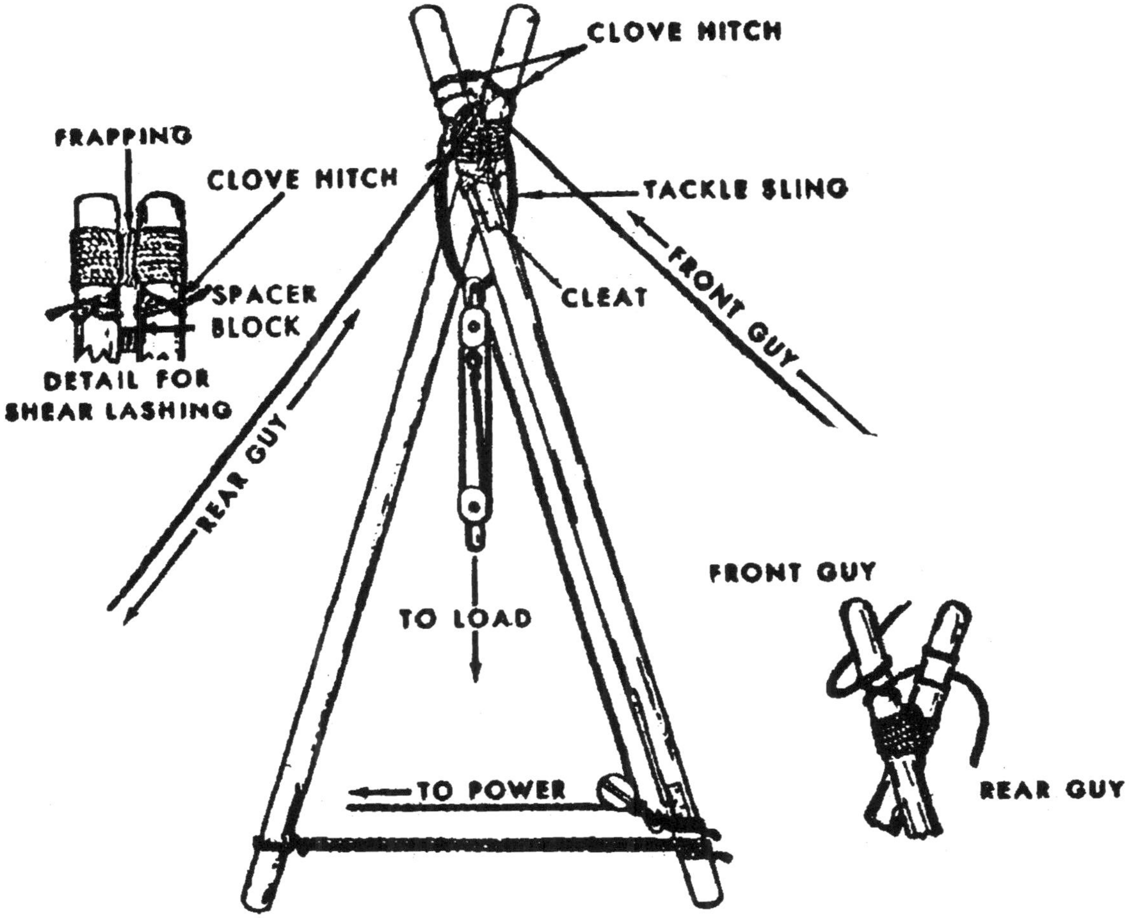

Figure 5-6. Lashing for shears.

60, because of the tendency of the legs to bend rather than to act as columns. For light work, shears can be improvised from two planks or light poles bolted together and reinforced by a small lashing at the intersection of the legs.

a. Rigging. In erection, the spread of the legs should equal about one-half the height of the shears. The maximum allowable drift (inclination) is 45°. Tackle blocks and guys for shears are essential. The guy ropes can be secured to firm posts or trees with a turn of the rope so that the length of the guys can be adjusted easily. The procedure is as follows:

(1) Lay two timbers together on the ground in line with the guys, with the butt ends pointing toward the back guy and close to the point of erection.

(2) Place a large block under the tops of the legs just below the point of lashing (fig. 5-6), and insert a small spacer block between the tops at the same point. The separation between the legs at this point should be equal to one-third the diameter on one leg, to make handling of the lashing easier.

(3) With sufficient 1-inch rope for 14 turns around both legs, make a clove hitch (fig. 5-6) around one spar, and take 8 turns around both legs above the clove hitch. Wrap the turns tightly so that the lashing is made smooth and without kinks.

(4) Finish the lashing by taking two frapping turns around the lashing between the legs and securing the end of the rope to the other leg just below the lashing. For handling heavy

Figure 5–7. Erecting shears.

loads the number of lashing turns is increased.

b. Erecting. Holes should be dug at the points where the legs of the shears are to stand. In case of placement on rocky ground, the base for the shears should be level. The legs of the shears should be crossed and the butts placed at the edges of the holes. With a short length of rope, make two turns over the cross at the top of the shears and tie the rope together to form a sling. Be sure to have the sling bearing against the spars and *not* on the shears lashing entirely. The procedure is as follows:

(1) Reeve a set of blocks and place the hook of the upper block through the sling. Secure the sling in the hook by mousing. Fasten the lower block to one of the legs near the butt, so that it will be in a convenient position when the shears have been raised, but will be out of the way during erection.

(2) If the shears are to be used on heavy lifts, another tackle is rigged in the back guy near its anchorage. The two guys should be secured to the top of the shears with clove hitches to legs opposite their anchorages above the lashing.

(3) Several men (depending on the size of the shears) should lift the top end of the shear

Figure 5–8. Hoisting with shears.

legs and "walk" them up by hand until the tackle on the rear guyline can take effect (fig. 5–7). After this, the shear legs can be raised into final position by hauling in on the tackle. Secure the front guyline to its anchorage before raising the shear legs and keep a slight tension on this line to control movement.

(4) The legs should be kept from spreading by connecting them with rope, chain, or boards. It may be necessary, under some conditions, to anchor each leg of the shears during erection to keep the legs from sliding in the wrong direction.

c. Operating. The rear guy is a very important part of the shears rigging, as it is under a considerable strain when hoisting. In order to avoid guyline failure, guylines should be designed according to the principles discussed in chapter 4, section II. The front guy has very little strain on it and is used mainly to aid in adjusting the drift and to steady the top of the shears when hoisting or placing the load. It may be necessary to rig a tackle in the rear guy for handling heavy loads. In operation, the drift (inclination of the shears) desired is set by adjustment of the rear guy, but this should not be done while a load is on the shears. For handling light loads, the fall line of the tackle of the shears can be led straight out of the upper block. When heavy loads are handled, it will be necessary to lash a snatch block (fig. 5-8) near the base of one of the shear legs to act as a leading block. The fall line should be run through the leading block to a hand- or power-operated winch for heavy loads.

5-4. Boom Derrick

A boom derrick is a lifting device which incorporates the advantages of a gin pole and the long horizontal reach of a boom. The boom derrick may be used to lift and swing medium size loads in a 90° arc on either side of the resting position of the boom, for a total swing of 180°. When a boom derrick is employed in lifting heavy loads, it must be set on a turnplate or turnwheel to allow the mast and boom to swing as a unit. A mast is a gin pole used with a boom. The mast can swing more than 180 degrees when it is set on a turnplate or turnwheel.

a. Rigging. For hoisting medium loads, a boom may be rigged to swing independently of the pole. Care must be taken to insure the safety of those using the installation, and it should be used only temporarily or where time does not permit a more stable installation. When using a boom on a gin pole, more stress is placed on the rear guy, and therefore a stronger guy is necessary. In case larger rope is not at hand, a set of tackle reeved with the same size rope used in the hoisting tackle can be used as a guyline by extending the tackle from the top of the gin pole to the anchorage. The block attached to the gin pole should be lashed at the point where the other guys are tied and in the same manner. The procedure is as follows:

(1) Rig a gin pole as described in paragraph 5-1*a*, but lash another block about 2 feet below the tackle lashing at the top of the pole (fig. 5-9). Reeve the tackle so that the fall line comes from the traveling block instead of the standing block. Attach the traveling block to the top end of the boom after the gin pole is erected.

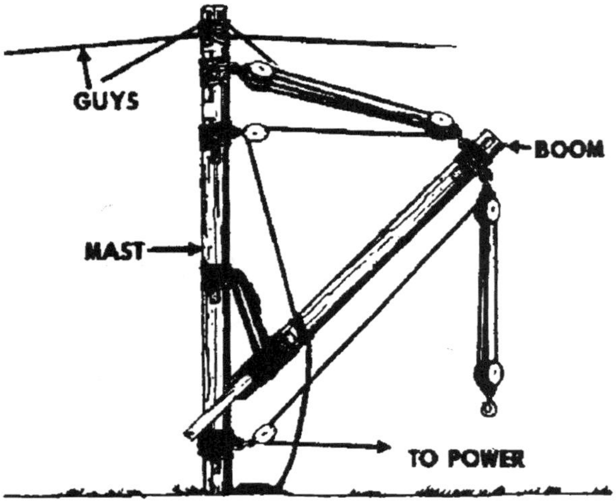

Figure 5-9. Rigging a boom on a gin pole.

(2) Erect the gin pole in the manner described in paragraph 5-1*b*, but pass the fall line of the tackle through the extra block at the top of the pole before erection to increase the mechanical advantage of the tackle system.

(3) Select a boom with the same diameter and not more than two-thirds as long as the gin pole. Spike two boards (fig. 5-9) to the butt end of the boom and lash them with rope, making a fork. The lashing should be made with a minimum of sixteen turns and tied off with a square knot. Drive wedges (fig. 5-9) under the lashing next to the cleats to help make the fork more secure.

(4) Spike cleats to the mast about 4 feet above the resting place of the boom and pla

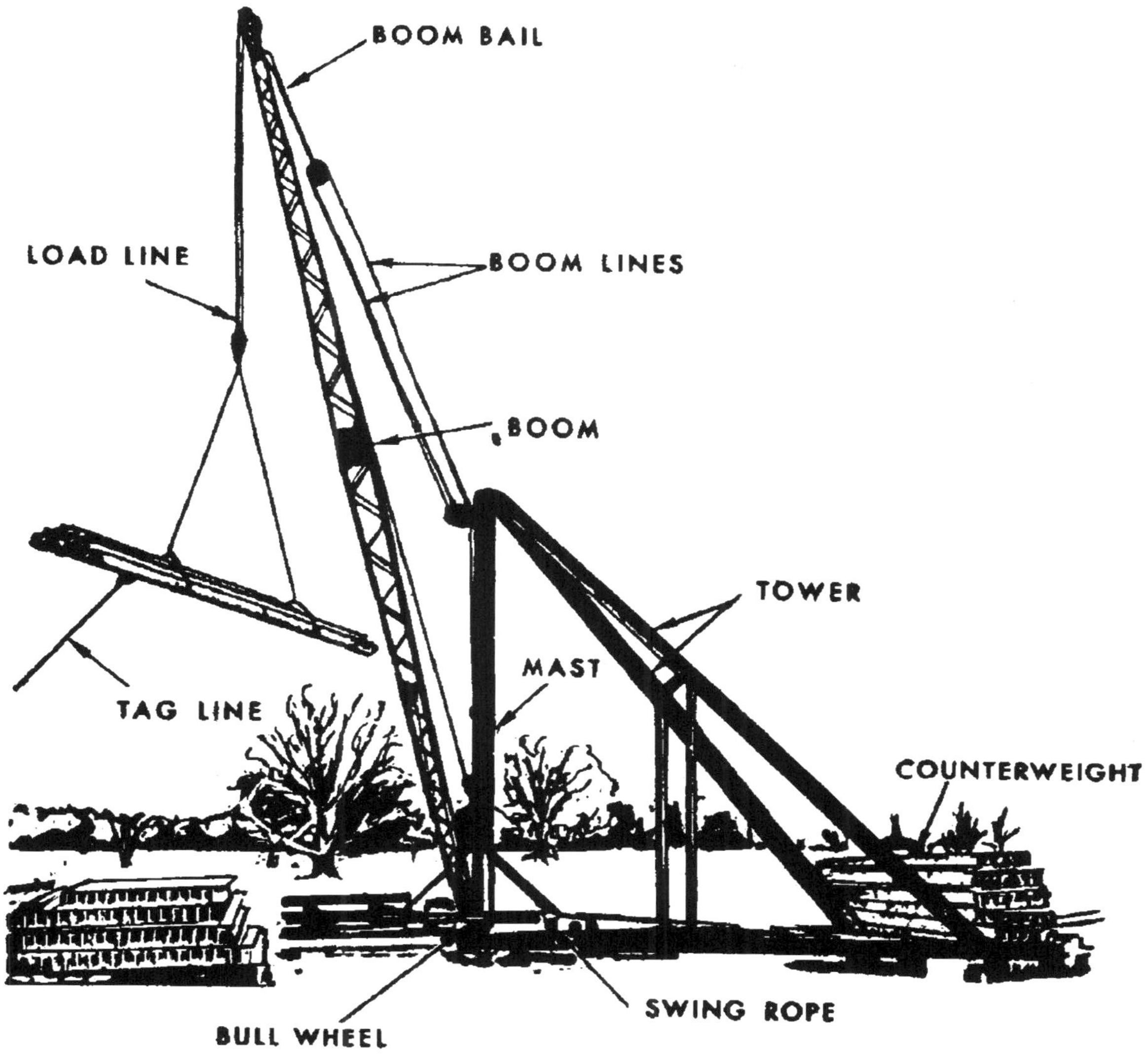

Figure 5–10. Four-ton stiff leg derrick.

another block lashing just above these cleats. This block lashing will support the butt of the boom. If a separate tackle system is rigged up to support the butt of the boom, an additional block lashing should be placed on the boom just below the larger lashing to secure the running block of the tackle system.

(5) If the boom is light enough, manpower may be used to lift the boom in place on the mast through the sling which will support it. The sling consists of 2 turns of rope with the ends tied together with a square knot. The sling should pass through the center 4 turns of the block lashing on the mast and should cradle the boom. On heavier booms, the tackle system on the top of the mast can be used to raise the butt of the boom to the desired position onto the mast.

(6) Lash the traveling block of the gin pole tackle to the top end of the boom as described in paragraph 5–1*a*, and lash the standing block of the boom tackle at the same point. Reeve the boom tackle so that the fall line comes from the standing block and passes

through the block at the base of the gin pole. The use of the leading block on this fall line is optional, but when handling heavy loads, more power may be applied to a horizontal line leading from the block with less strain on the boom and guys.

b. Erecting. The boom is raised into position when the rigging is finished. When working with heavy loads, the base of the boom should rest on the ground at the base of the pole. A more horizontal position may be used when working with light loads. In no case should the boom bear against any part of the upper two-thirds of the mast.

c. Operating. A boom on a gin pole provides a convenient means for loading and unloading trucks or flatcars when the base of the gin pole cannot be set close to the object to be lifted. It is used also on docks and piers for unloading boats and barges. The boom is swung by pushing directly on the load or by pulling the load with bridle lines or tag lines. The angle of the boom to the mast is adjusted by hauling on the fall line of the mast tackle. The load is raised or lowered by hauling on the fall line of the boom tackle. A leading block (snatch block) is usually placed at the base of the gin pole. The fall line of the boom tackle is led through this leading block to a hand- or power-operated winch for the actual hoisting of the load.

5–5. Stiff Leg Derrick

The mast of a stiff leg derrick is held in the vertical position by two rigid, inclined struts connected to the top of the mast. The struts are spread 60° to 90° to provide support in two directions and are attached to sills extending from the bottom of the mast. The mast is mounted on vertical pins. The mast and boom can swing through an arc of about 270°. The tackles for hoisting the load and raising the boom are similar to those used with the boom and gin pole (par. 5–4*a*).

a. Operating. A stiff leg derrick equipped with a long boom is suitable for yard use for unloading and transferring material whenever continuous operations are carried on within reach of its boom. When used on a bridge deck these derricks must be moved on rollers. They are sometimes used in multistoried buildings surmounted by towers to hoist material to the roof of the main building to supply guy derricks mounted on the tower. The stiff leg derrick also is used where guylines cannot be provided, as on the edge of a wharf or on a barge.

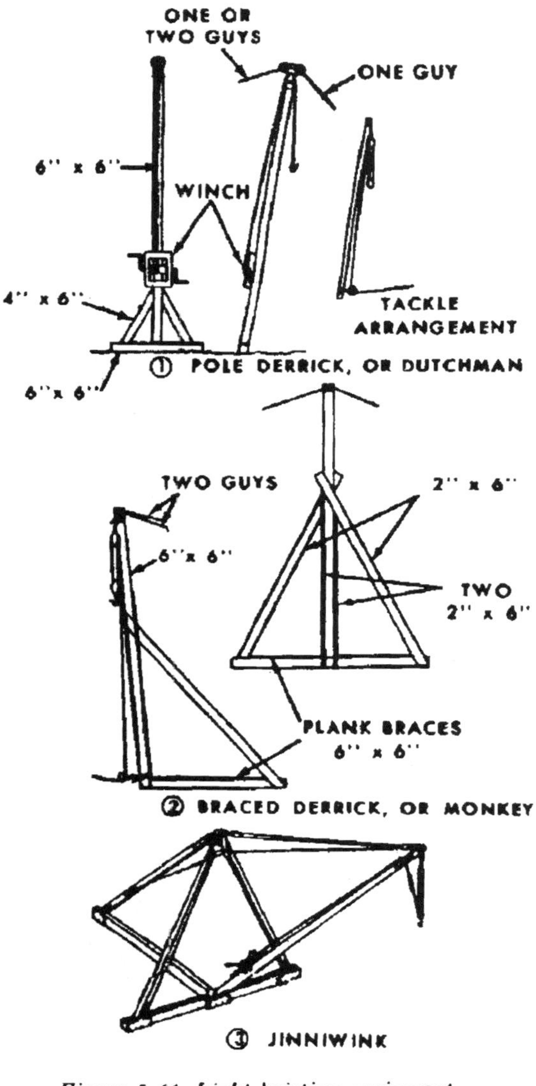

Figure 5–11. Light hoisting equipment.

b. Steel Derrick. Steel derricks of the stiff leg type are available to engineer troops in two sizes: 4-ton rated capacity (fig. 5–10) with a 28-foot radius, and a 30-ton rated capacity with a 38-foot radius, when properly counterweighted. Both derricks are erected on fixed bases. The 4-ton derrick, including a skid-mounted double-drum gasoline-engine-driven hoist, weighs 7 tons and occupies a space 20 feet square. The 30-ton derrick, including a skid-mounted double-drum hoist, weighs approximately 22 tons and occupies a space 29 feet square.

5–6. Light Hoisting Equipment

Extended construction projects generally involve the erection of numerous light members as well as the heavy main members. Progress can be more rapid if light members are raised by hand or by light hoisting equipment, allowing the heavy hoisting equipment to move ahead with the erection of the main members. Very light members can be raised into place by two men using manila handlines. Where handlines are inadequate or where members must be raised above the working level, light hoisting equipment should be used. Many types of hoisting equipment for lifting light loads have been devised. Those discussed here are only typical examples which can be constructed easily in the field and moved readily about the job.

a. Pole Derrick. The improved pole derrick, called a "dutchman" (①, fig. 5–11), is essentially a gin pole constructed with a sill and knee braces at the bottom. It is usually installed with guys at the front and back. It is effective for lifting loads of 2 tons and, because of its light weight and few guys, is readily moved from place to place by a small squad.

b. Brace Derrick. The braced derrick, known as a "monkey" (②, fig. 5–10), is very useful for filling in heavy members behind the regular erection equipment. Two back guys are usually employed when lifting heavy loads, although light members may be lifted without them. Power is furnished by a hand- or power-driven hoist. The construction of the base of the monkey permits it to be anchored to the structure by lashings to resist the pull of the lead line on the snatch block at the foot of the mast.

c. Jinniwink Derrick. This derrick (②, fig. 5–11) is suitable for lifting loads weighing 5 tons. Hand-powered jinniwinks are rigged preferably with manila rope. Those operated by a power-driven hoist should be rigged with wire rope. The jinniwink is lashed down to the structural frame at both the front sill and tail sill to prevent the tail sill from rising when a load is lifted.

Section II. SKIDS, ROLLERS, AND JACKS

5–7. Introduction

Skids, rollers, and jacks are used to move heavy loads. Cribbing or blocking is often necessary as a safety measure to keep an object in position or to prevent accidents to personnel that work under or near these heavy objects. Cribbing is formed by piling timbers in tiers, with the tiers alternating in direction (fig. 5–12), to support a heavy weight at a height greater than blocking would provide. A firm and level foundation for cribbing is essential, and the bottom timbers should rest firmly and evenly on the ground. Blocking used as a foundation for jacks should be sound and large enough to carry the load. The timbers should be dry and free from grease, and placed firmly on the ground so that the pressure is evenly distributed.

5–8. Skids

Timber skids may be placed longitudinally under heavy loads (fig. 5–13) to distribute the weight over a greater area, to make a smooth surface for skidding equipment, or to provide a runway surface when rollers are used. Oak planks 2 inches thick and about 15 feet long make satisfactory skids for most operations. The angle of the skids must be kept low to prevent the load from drifting or getting out of control. Grease may be used on skids when horizontal movement only is involved, but in most circumstances greasing is dangerous as it may cause the load to drift sideways suddenly.

5–9. Rollers

Hardwood or pipe rollers can be used over skids for moving very heavy loads into position. Skids are placed under the rollers to provide a smooth, continuous surface for the rollers. The rollers must be smooth and round and should be long enough to pass completely under the load being moved. The load should be supported on longitudinal wooden members to provide a smooth upper surface for the rollers to move on. The skids placed underneath the rollers must form continuous support. Ordinarily, four to six rollers (fig. 5–13) are placed under the load to be moved. Several

Figure 5-12. Timber cribbing.

rollers are placed in front of the load and the load is rolled slowly forward unto the rollers. As the load passes, rollers are left clear behind the load and are picked up and placed in front of the load so that there is a continuous path of rollers. In making a turn with a load on rollers, the front rollers must be inclined slightly in the direction of the turn and the rear rollers in the opposite direction. This inclination of the rollers may be made by striking them sharply with a sledge. For moving lighter loads, rollers can be made up and set on axles in side beams as a semipermanent conveyor. Permanent metal roller conveyors (fig. 5-14) are available. They are usually made in sections.

5-10. Jacks

a. Use. In order to place cribbing, skids, or rollers, it is often necessary to lift and lower the load for a short distance. Jacks are used for this purpose. Jacks are used also for precision placement of heavy loads, such as bridge spans. A number of different styles of jacks are available, but only heavy duty hydraulic or screw type jacks should be used. The number of jacks used will depend on the weight of the load and the rated capacity of the jacks. Be certain that the jacks are provided with a solid footing, preferably wooden blocking. Cribbing is frequently used in lifting loads by jacking stages (fig. 5-15). The procedure requires blocking under the jacks, raising of the object tc the maximum height of the jacks to permit cribbing to be put directly under the load, and the lowering of the load onto the cribbing.

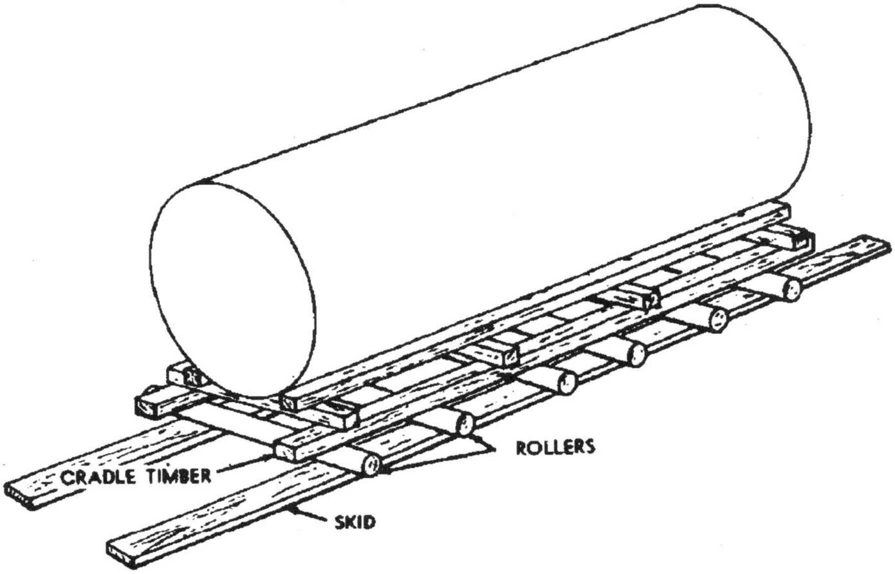

Figure 5-13. Using skids and rollers.

Figure 5-14. Metal conveyors.

This process is repeated as many times as necessary to lift the load to the desired height.

b. Types. Jacks are available in capacities from 5 to 100 tons (fig. 5-16). Small capacity jacks are operated through a rack bar or screw, while those of large capacity are usually operated hydraulically.

Figure 5-15. Jacking loads by stages.

(1) *Ratchet lever jacks.* The ratchet lever jack (①, fig. 5-16), available to engineer troops as part of panel bridge equipment, is a rack-bar jack which has a rated capacity of 15 tons. It has a foot lift by which loads close to its base can be engaged. The foot capacity is 7½ tons.

(2) *Steamboat ratchets.* Steamboat ratchets (sometimes called pushing-and-pulling jacks) (②, fig. 5-16) are ratchet screw jacks of 10-ton rated capacity with end fittings which permit pulling parts together or pushing them apart. Their principal uses are for tightening lines or lashings and for spreading or bracing parts in bridge construction.

(3) *Screw jacks.* Screw jacks (③, fig. 5-16) have a rated capacity of 12 tons. They are about 13 inches high when closed and have a safe rise of at least 7 inches. These jacks are issued with the pioneer set and can be used for general purposes, including steel erection.

(4) *Hydraulic jacks.* Hydraulic jacks (④, fig. 5-16) are available in class IV supplies in capacities up to 100 tons. Loads normally encountered by engineer troops do not require large capacity hydraulic jacks. Those supplied with the squad pioneer set are 11 inches high, have a rated capacity of 12 tons, and a rise of at least 5¼ inches. They are large enough for usual construction needs.

1 Ratchet lever jack with foot lift　　2 Steamboat ratchet　　3 Screw jack　　4 Hydraulic jack

Figure 5–16. Mechanical and hydraulic jacks.

CHAPTER 6

LADDERS AND SCAFFOLDING

Section I. LADDERS

6–1. Introduction

Several types of ladders are available for construction work, including extension ladders, pushup ladders, and straight ladders (fig. 6–1). All three types of ladders are available in both metal and wood. Ladders should always be inspected before they are used. A ladder with parts missing, with bent or cracked sides or rungs, and those made with faulty material should be condemned. Badly worn and weathered ladders and wooden ladders with rotten spots should not be used because they are subject to breaking and can cause a serious accident. Ladders with rough spots, such as protruding metal fastenings, screws, and nails should be repaired or reconstructed to prevent injuries.

Figure 6–1. Types of ladders.

6–2. Extension Ladder

Occasionally sections of an extension ladder are used separately. When this is done, the upper section of the ladder must be used upside down so that the rung missing at the locks will be at the top of the ladder where it is less liable to cause an accident. In selecting an extension ladder for a particular job, it should be remembered that this type of ladder is designated by its nominal length, which is the sum of the lengths of the sections. The usable length of the ladder is 3 to 10 feet less than the nominal length due to the overlap of the sections. This overlap is 3 feet on ladders up to and including 36 feet, 4 feet on 40 to 44 feet ladders, and 5 to 10 feet on longer ones. Extension ladders and pushup ladders are placed against the wall in much the same manner with the extension, or pushup, portion lowered. When the lower section of his ladder is nearly in the final position, the upper portion is pulled out away from the building until it stands nearly vertical but leaning slightly toward the building. While the ladder is held in this position, one man hauls down on the rope fastened to the extension section, pulling it upward. No attempt should be made to raise the extension section to its full extension on the first pull. It is less difficult to pull the section up in easy stages, checking the height of the ladder at intervals in order to determine the correct height. The extension section should be on the side of the ladder toward the building to lessen the danger of the user slipping at the point where the two sections join.

6–3. Straight Ladders

Straight ladders are placed on the ground beside the wall which they are intended to scale, and at right angles to the wall. One man should stand at the foot of the ladder to prevent the ladder from kicking backward. A second man (or men) grasp the ladder part way toward the top and raise it from the ground. As the ladder is raised it is "walked" toward the building, and the men keep moving toward the foot of the ladder to grasp new holds. When the ladder is in final position, the bottom of the ladder should be checked to make certain it has a firm footing. If the ground is soft, or if the ladder does not rest squarely on both bottom legs, a board may be placed under the feet. If the bottom of the ladder has antislip surfaces, these surfaces should be checked before placing the ladder to be certain they are not coated with mud or debris. The ladder should be placed at a safe angle against the wall. A good rule is to place the base of the ladder about one-fourth as far out from the upper support as the length of the ladder (fig. 6–2). The upper end of the ladder should not extend more than 2 feet above the upper support, and not so far below the working area to be dangerous to move from the top of the ladder to the wall. The upper end of a ladder should always be lashed to the structure with wire or fiber rope to prevent it from skidding sideways or overturning while in use.

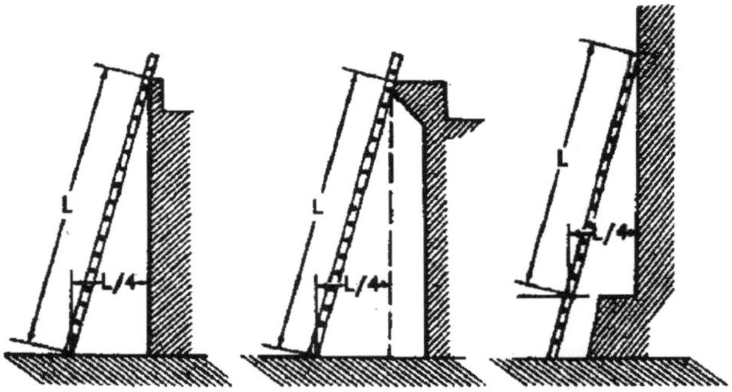

Figure 6–2. Correct angles for ladders.

Section II. SCAFFOLDING

6–4. Introduction

Construction jobs may require the use of several kinds of scaffolds to permit easy working procedures. Scaffolding may range from individual planks placed on structural members of the building to involved patent scaffolding. Scaffold planks are placed as a decking over swinging scaffolds, suspended scaffolds, needle beams, and built-up independent scaffolding. Scaffold planks are of various sizes, including 2 x 9 inches x 13 feet, 2 x 10 inches x 16 feet, and 2 x 12 inches x 16 feet. Scaffold planks 3 inches thick may be needed for platforms that must hold heavy loads or withstand movements. Planks with holes or splits are not suitable for scaffolding if the diameter of the hole is more than 1 inch or the split extends more than 3 inches in from the end. Three-inch planks should be used to build the temporary floor used for construction of steel buildings because of the possibility that a heavy steel member might be rested temporarily on the planks. Single scaffold planks may be laid across beams of upper floors (fig. 6–3) or roofs to form working areas or runways. Each plank should run from beam to beam, with not more than a few inches of any plank projecting beyond the end of the supporting beam. Overhangs are dangerous because men may step on them and over balance with the scaffold plank. When the planking is laid continuously, as in a runway, the planks should be laid so that their ends overlap. Single plank runs can be staggered so that each plank is offset with reference to the next plank in the run. It is advisable to use two layers of planking on large working areas to increase the freedom of movement.

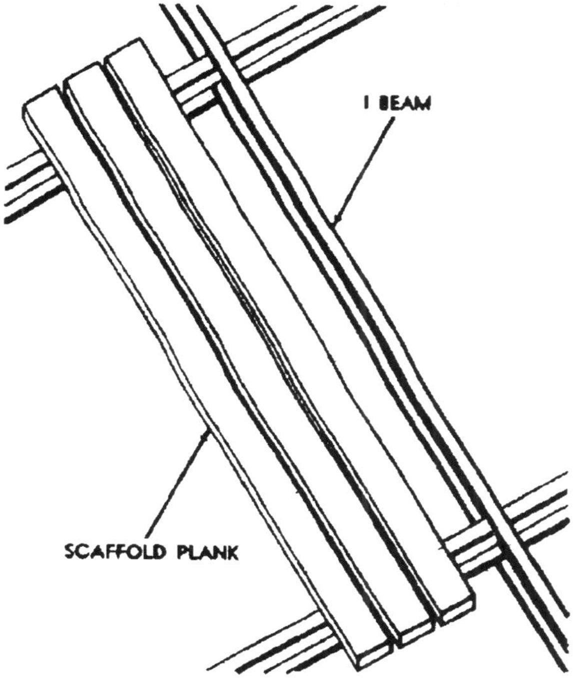

Figure 6–3. Scaffold planks in place.

6–5. Types of Scaffolds

a. Swinging Scaffolds. The swinging, single plank, or platform type of scaffold must always be secured to the building or structure to prevent it from moving away and causing the mend to fall. Where swinging scaffolds are suspended adjacent to each other, planks should never be placed so as to form a bridge between them.

(1) *Single plank scaffold.* A single scaffold plank (fig. 6–4) may be swung over the edge of a building with two ropes by using a scaffold hitch (fig. 2–28) at each end. A tackle may be inserted in place of ropes for lowering and hoisting. This type of swinging scaffold is suitable for one man.

(2) *Platform scaffold.* The swinging platform scaffold (fig. 6–5) consists of a frame similar in appearance to a ladder with a decking of wood slats. It is supported near each end by a steel stirrup to which the lower block of a set of manila rope falls is attached. The scaffold is supported by hooks or anchors on the roof of a structure. The fall line of the tackle must be secured to a member of the scaffold when in final position to prevent it from falling.

b. Suspended Scaffolds. Suspended scaffolds usually are supported on outriggers at the roof and are heavier than swinging scaffolds. From each outrigger, cables lead to hand winches on the scaffold. This type of scaffold is raised or lowered by operating the hand winches, which must contain a locking device. The scaffold may be made up in almost any width up to about 6 feet, and may be 12 feet long, depending on the size of the putlogs, or longitudinal supports, under the scaffold. A light roof may be included on this type of scaffold to protect the men from falling debris.

c. Needle Beam Scaffold. This type of scaffold is used only for temporary jobs. No material should be stored on this scaffold. In needle beam scaffolding, two 4- x 6-inch, or similar size, timbers are suspended by ropes. A decking of 2-inch scaffold plank is placed across the needle beams, which should be placed about 10 feet apart. Needle beam scaffolding (fig. 6–6) is used frequently by riveting gangs working on steel structures because of the frequent changes of location necessary and its adaptability to different situations. A scaffold hitch is used in the rope supporting the needle beams to prevent them from rolling or turning over. The hanging lines are usually of 1¼-inch manila rope. The rope is hitched to the needle beam, carried up over a structural beam or other support, and then down again under the needle beam so the latter has a complete loop of rope under it. The rope is then passed over the support again and fastened around itself by two half hitches.

d. Double Pole Built-Up Scaffold. The double pole built-up scaffold (steel or wood), sometimes called the independent scaffold, is completely independent of the main structure. Several types of patent independent scaffolding are available for simple and rapid erection (fig. 6–7). The scaffolding can be built up from wood members if necessary. The scaffold uprights are braced with diagonal members and the working level is covered with a platform of planks. All bracing must form triangles and the base of each column requires adequate footing plates for the bearing area on the ground. The patented steel scaffolding is usually erected by placing the two uprights on the ground and inserting the diagonal members.

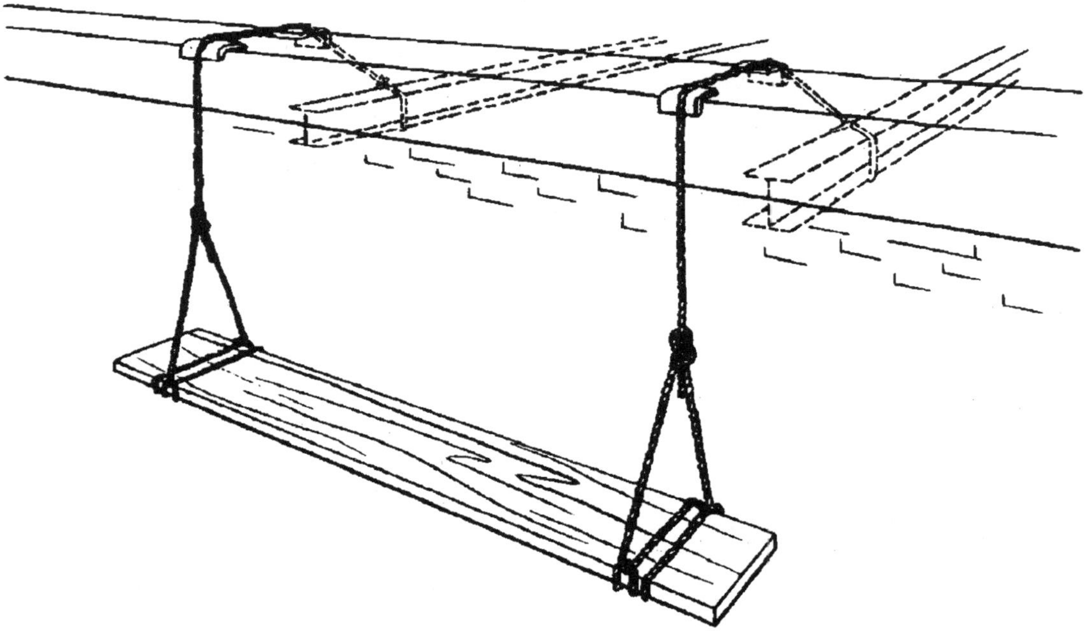

Figure 6-4. Single swinging plank scaffold.

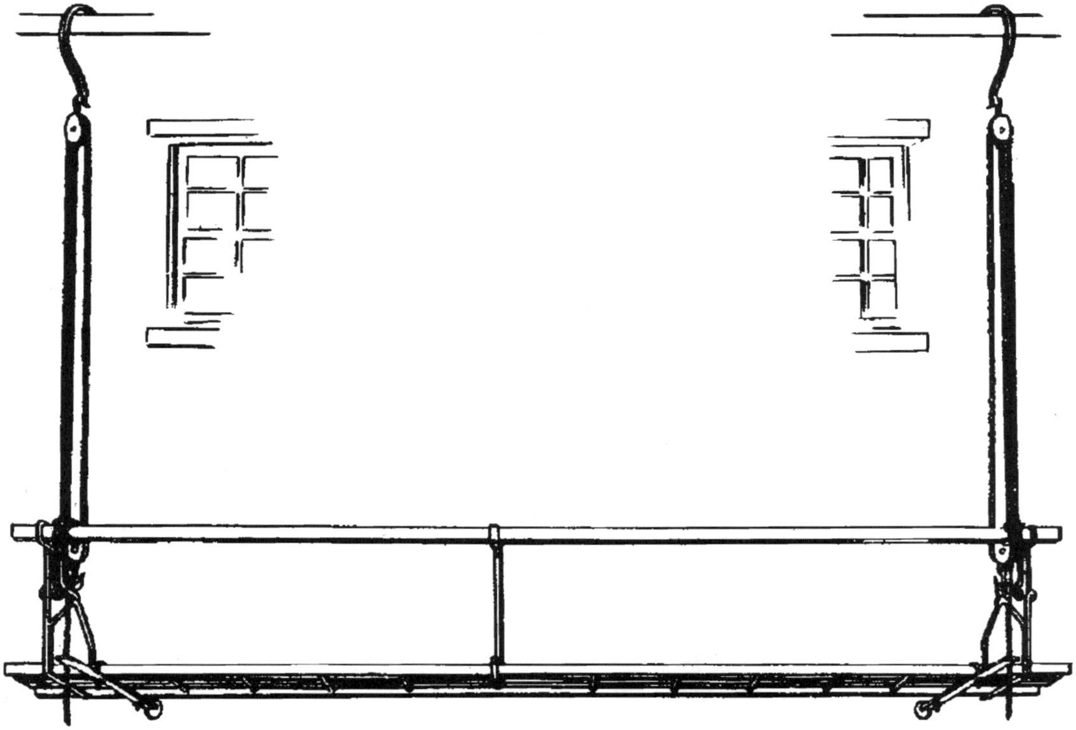

Figure 6-5. Swinging platform scaffold.

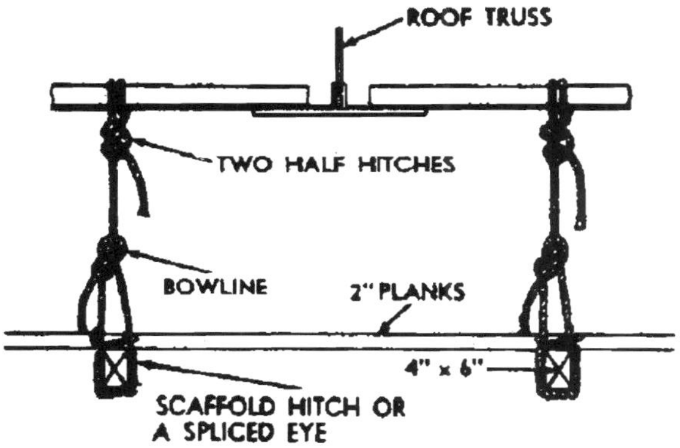

Figure 6–6. Needle beam scaffold.

The diagonal members have end fittings which permit rapid locking in position. The first tier is set on steel bases on the ground. A second tier is placed in the same manner on the first tier, with the bottom of each upright locked to the top of the lower tier. A third and fourth upright can be placed on the ground level and locked to the first set with diagonal bracing. The scaffolding can be built as high as desired, but high scaffolding should be tied in to the main structure.

6–6. Boatswain's Chair

The boatswain's chair can be made up in several forms, but it generally consists of a sling for supporting one man.

a. Rope Chair. A rope boatswain's chair (fig. 6–8) can be made up by using a double bowline and a rolling hitch. One man can operate the rope seat to lower himself by releasing the grip of the rolling hitch. A slight twist with the hand on the hitch permits the suspension line to slip through it, but when the hand pressure on the hitch is released, the hitch will hold firmly.

b. Rope Chair With Seat. If the rope boatswain's chair must be used to support a man at work for some time, the rope may cause considerable discomfort. A notched board (fig. 6–9) inserted through the two leg loops will provide a comfortable seat. The loop formed as the running end to make the double bowline will still provide a back support, and the rolling hitch can still be used to lower the boatswain's chair.

c. Boatswain's Chair With Tackle. The boatswain's chair is supported by a four part rope tackle (fig. 6–10), two double blocks. One man can raise or lower himself, or be assisted by a man on the ground. When working alone the fall line is attached to the lines between the seat and the traveling block with a rolling hitch. As a safety precaution, a figure eight knot should be tied after the rolling hitch to prevent accidental untying.

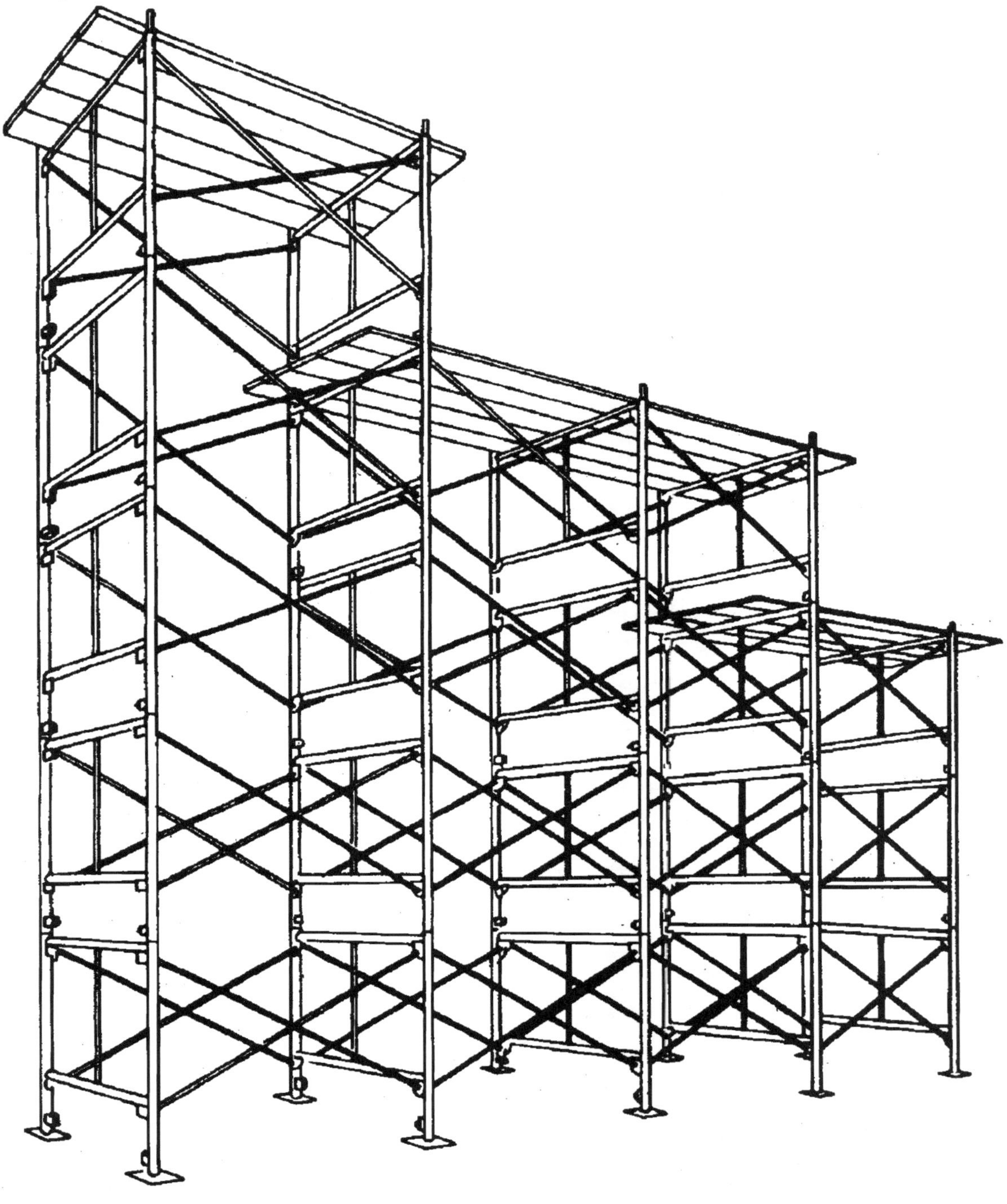

Figure 6–7. Independent scaffolding.

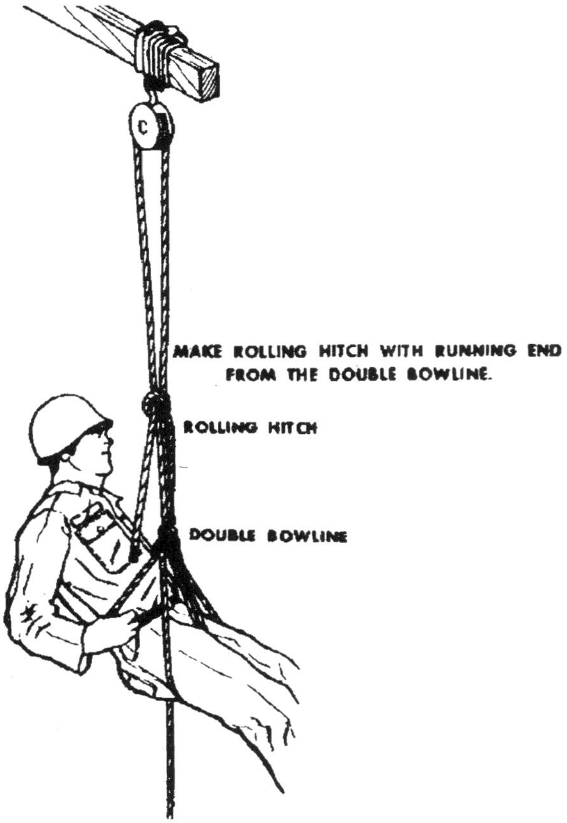

Figure 6–8. Boatswain's chair.

Figure 6–9. Boatswain's chair with seat.

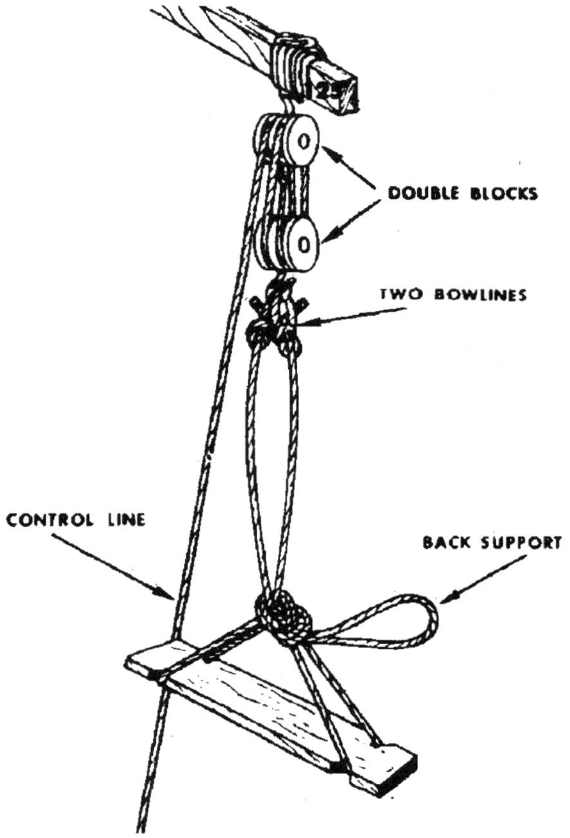

Figure 6–10. Boatswain's chair with tackle.

APPENDIX A

REFERENCES

FM 5–13	The Engineer Soldier's Handbook
FM 5–34	Engineer Field Data
FM 5–35	Engineer Reference and Logistical Data
FM 20–22	Vehicle Recovery Operations
FM 55–15	Transportation Reference Data
TB ENG 300-series	Air Movement Instructions
TM 5–270	Cableways, Tramways, and Suspension Bridges
TM 5–744	Structural Steelwork
TM 10–500-series	Airdrop of Supplies and Equipment
TM 57–210	Air Movement of Troops and Equipment

APPENDIX B

TABLES OF USEFUL INFORMATION

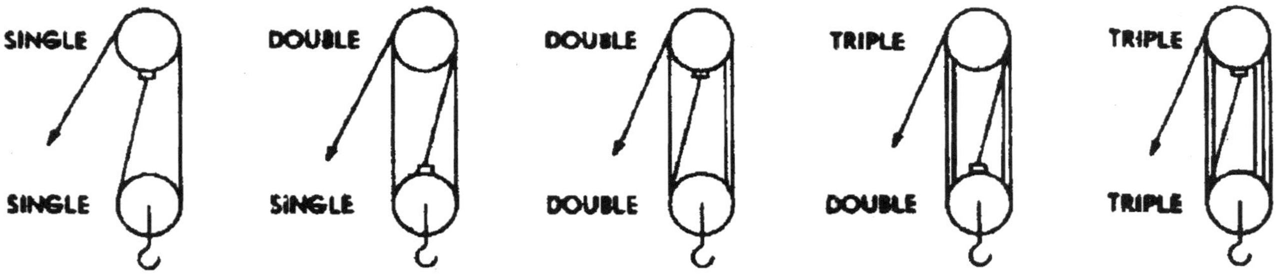

Load to be lifted (tons)		Total number of sheaves in blocks				
		2 (2-single blocks)	3 (1-single 1-double)	4 (2-double blocks)	5 (1-double 1-triple)	6 (2-triple blocks)
1/2	Smallest permissible rope diameter (inch)	1/2	7/16	3/8	3/8	3/8
	Lead line pull (lbs)	540	380	300	250	220
1	Rope diameter	3/4	5/8	1/2	1/2	1/2
	Pull	1, 100	760	600	500	440
1 1/2	Rope diameter	7/8	3/4	5/8	5/8	1/2
	Pull	1, 600	1, 100	900	750	660
2	Rope diameter	1 1/8	7/8	3/4	5/8	5/8
	Pull	2, 200	1, 600	1, 200	1, 000	880
3	Rope diameter	5/16	1 1/8	1	7/8	3/4
	Pull	3, 300	2, 300	1, 800	1, 500	1, 300
4	Rope diameter	1 1/2	1 1/4	1 1/8	1	1
	Pull	4, 400	3, 000	2, 400	2, 000	1, 800
6	Rope diameter	----------	1 1/2	1 5/16	1 1/4	1 1/8
	Pull	----------	4, 500	3, 600	3, 000	2, 600
8	Rope diameter	----------	----------	1 5/8	1 1/2	1 5/16
	Pull	----------	----------	4, 800	4, 000	3, 500

Note. Permissible rope diameters are for new ropes used under favorable conditions as rope ages or deteriorates increase factor of safety progressively to 8, when selecting rope size. Lead line pull is not affected by age or condition.

Table B–1. Simple Block and Tackle Rigging for Manila Rope (Factor of Safety 5).

Table B-2. Simple Block and Tackle Rigging for Plow Steel Wire Rope
(Factor of Safety 6)

Load to be lifted (tons)		Total number of sheaves in blocks				
		2 (2-single blocks)	3 (1-single 1-double)	4 (2-double blocks)	5 (1-double 1-triple)	6 (2-triple blocks)
1	Smallest permissible rope diameter (inch)	⅜	⅜	⅜	⅜	⅜
	Lead line pull (lbs)	1,000	720	560	460	400
2	Rope	½	⅜	⅜	⅜	⅜
	Pull	2,100	1,400	1,100	920	800
4	Rope	⅝	½	½	⅜	⅜
	Pull	4,200	2,900	2,200	1,800	1,600
6	Rope	¾	⅝	⅝	½	½
	Pull	6,200	4,300	3,400	2,800	2,400
8	Rope	⅞	¾	⅝	⅝	⅝
	Pull	8,300	5,800	4,500	3,700	3,200
10	Rope	1	⅞	¾	⅝	⅝
	Pull	10,400	7,200	5,600	4,600	4,000
15	Rope	1⅛	1	⅞	¾	¾
	Pull	15,600	10,800	8,400	6,900	6,000
20	Rope	1½	1⅛	1	⅞	⅞
	Pull	20,800	14,400	11,200	9,200	8,000

Table B-3. Recommended Sizes of Tackle Blocks

Wire rope		Manila rope	
Rope diameter (inches)	Outside-diameter of sheave (inches)	Rope diameter (inches)	Length of shell (inches)
⅜	6–8	½	4
½	8–10	⅝	6
⅝	10–12	¾	6–7
¾	12–16	⅞	7–8
⅞	14–18	1	8–10
1	14–20	1⅛	8–10
		1¼	10–12
		1½	12–14
		1¾	14–16

Note. Largest diameter of sheave for a given size of rope is preferred, when available, except that for 6 x 37 wire rope the smaller diameter of sheave is suitable.

Table B-4. Bearing Capacity of Soils

General description	Condition	Safe allowable pressure (psf)
Fine grained soils: Clays, silts, very fine sands, or mixtures of these containing few coarse particles of sand or gravel. Classification: MH, CH, OH, ML, CL, OL.	Soft, unconsolidated, having high moisture content (mud).	1,000
	Stiff, partly consolidated, medium moisture content	4,000
	Hard, well consolidated, low moisture content (slightly damp to dry).	8,000
Sands and well-graded sandy soils, containing some silt and clay. Classification: SW, SC, SP, SF.	Loose, not confined	3,000
	Loose, confined	5,000
	Compact	10,000
Gravel and well-graded gravelly soils containing some sand, silt and clay. Cassification: GW, GC, GP.	Loose, not confined	4,000
	Loose, confined	6,000
	Compact	12,000
	Cemented sand and gravel	16,000
Rock	Poor quality rock, soft and fractured; also hardpan.	10,000
	Good quality; hard and solid	20,000

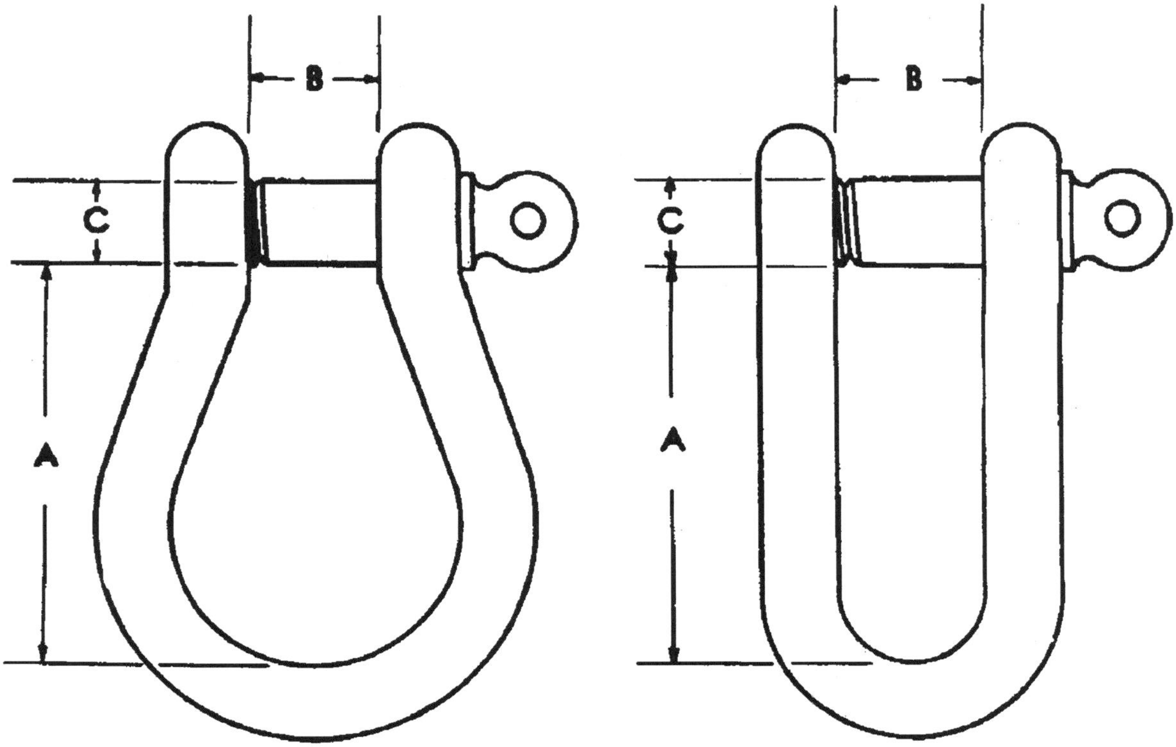

Size	Length of opening (in.) A	Distance between eyes (in.) B	Diameter of pin (in.) C	Safe load in pounds
¾	2⅞	1¼	⅞	6, 500
⅞	3¼	1⅜	1	8, 800
1	3⅝	1 11/16	1⅛	11, 000
1⅛	4¼	1⅞	1¼	13, 000
1¼	4¾	1⅞	1⅜	16, 000
1⅜	5¼	2¼	1½	19, 000
1½	5⅝	2¼	1⅝	23, 000
1¾	7	2⅞	2	35, 000
2	7¾	3⅛	2¼	42, 000

Table B–5. Safe Loads on Screw-Pin Shackles

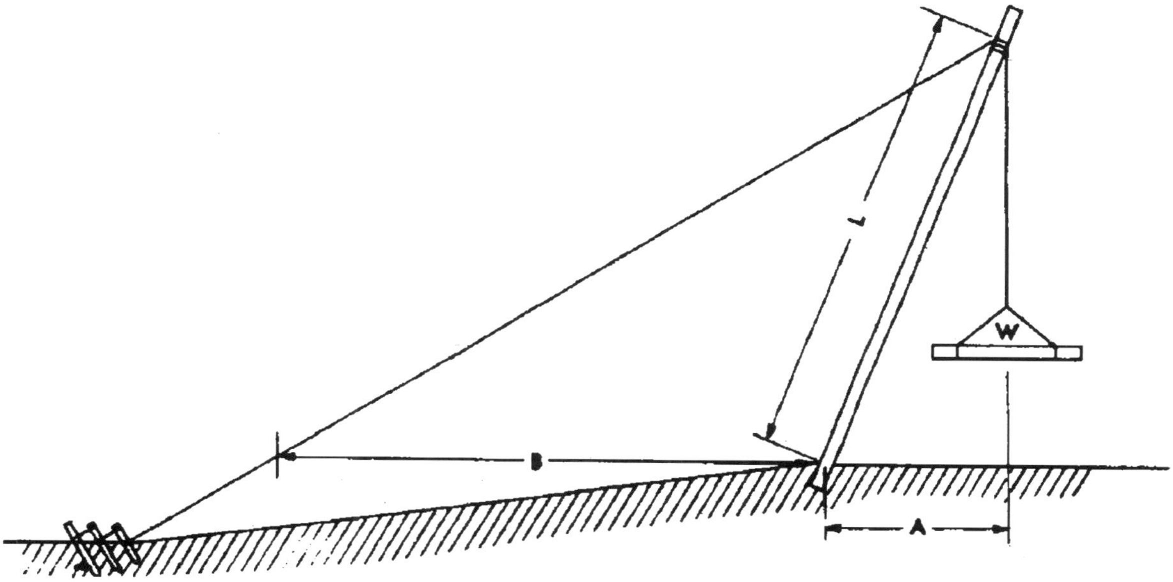

A	Stress (pounds) in guy for w=1,000 pounds				
	B=1/3L	B=1/2L	B=L	B=1 1/2L	B=2L
0	0	0	0	0	0
1/10L	230	180	150	130	120
1/8L	300	220	190	160	150
1/6L	400	300	260	220	200
1/4L	630	480	410	350	320
1/3L	890	680	580	480	440
	Stress (pounds) in spar for w=1,000 pounds				
0	1,000	1,000	1,000	1,000	1,000
1/10L	1,210	1,140	1,100	1,070	1,050
1/8L	1,260	1,180	1,140	1,090	1,070
1/6L	1,350	1,240	1,180	1,130	1,100
1/4L	1,550	1,380	1,290	1,210	1,160
1/3L	1,770	1,530	1,420	1,300	1,240

Key—W=weight to be lifted plus 1/2 the weight of pole.
A=Drift.
B=Horizontal distance from base of pole to guy.
L=Length of gin pole.

Table B-6. Stresses in Guys and Spars of Gin Poles

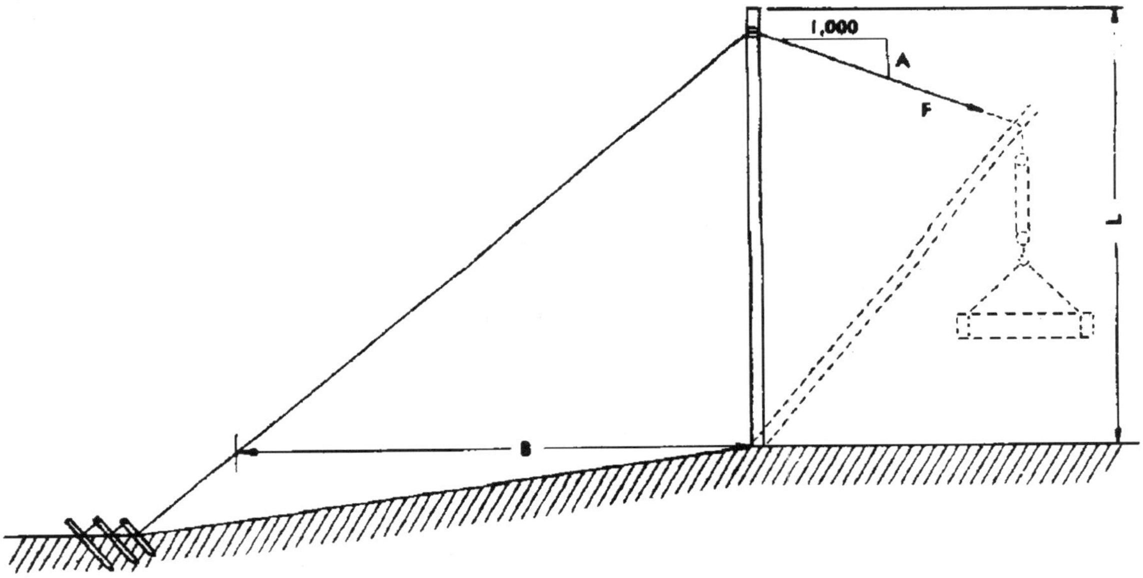

A	Stress (pounds) in guy for F=1,000 pounds				
	B=1/2L	B=3/4L	B=L	B=1 1/2L	B=2L
0	2, 240	1, 670	1, 420	1, 200	1, 120
0.50	2, 000	1, 490	1, 260	1, 080	1, 000
0.667	1, 860	1, 390	1, 180	1, 000	930
1.00	1, 570	1, 180	1, 000	850	790
1.33	1, 340	1, 000	850	720	670
2.00	1, 000	750	630	540	500
	Stress (pounds) in mast for F=1,000 pounds				
0	2, 000	1, 330	1, 000	670	500
0.50	2, 240	1, 640	1, 340	1, 040	900
0.667	2, 220	1, 660	1, 390	1, 110	970
1.00	2, 120	1, 650	1, 410	1, 180	1, 060
1.33	2, 000	1, 600	1, 400	1, 200	1, 100
2.00	1, 800	1, 490	1, 340	1, 190	1, 120

Key—F = Total force on boom lift falls.
A = Vertical distance for each unit of horizontal distance.
B = Horizontal distance from base of mast to guy.
L = Length of mast.

Table B-7. Stresses in Guys and Mast of Guy Derrick

INDEX

By Order of the Secretary of the Army:

WILLIAM C. WESTMORELAND,
General, United States Army,
Chief of Staff.

OFFICIAL:
KENNETH G. WICKHAM,
Major General, United States Army,
The Adjutant General.

Distribution:
Active Army:

DCSLOG (2)
USASA (2)
CNGB (5)
CofEngrs (5)
USACDCEA (5)
USAARTYBD (5)
USAARENBD (5)
USAIB (5)
USARADBD (5)
USAAESWBD (5)
USAAVNTBD (5)
USCONARC (5)
USAMECOM (5)
ARADCOM (2)
OS Maj Comd (10)
MDW (2)
Armies (5)
Corps (3)
Div (2)
Engr Bde (5)
Engr Gp (5)
Engr Bn (5)
Engr Co (5)
Svc Colleges (2)
Br Svc Sch (2) except
USAES (6000)
USAIS (50)
USAQMS (10)
PMS Sr Div Unit (2)
USMA (50)
Instl (5)
Gen Dep (OS) (5)
Engr Sec, Gen Dep (2)
Engr Dep (OS) (2)
EAMTMTS (2)
WAMTMTS (2)
MOTBA (2)
MOTBY (2)
MOTKI (2)
MOTSU (2)
Engr Cen (5)
Army Pic Cen (2)
USACOMZEUR (5)
USATCFLW (50)

NG: State AG (3); Units—Same as active Army except allowance is one (1) copy to each unit.
USAR: Same as active Army except allowance is one (1) copy to each unit.
For explanation of abbreviations used, see AR 320-50.

★ U.S. GOVERNMENT PRINTING OFFICE: 1979—280-762/488

This page intentionally left blank